New Neoliberalism and the Other

New Neoliberalism and the Other

Biopower, Anthropophagy, and Living Money

Giuseppe Cocco
Bruno Cava

LEXINGTON BOOKS
Lanham • Boulder • New York • London

Published by Lexington Books
An imprint of The Rowman & Littlefield Publishing Group, Inc.
4501 Forbes Boulevard, Suite 200, Lanham, Maryland 20706
www.rowman.com

Unit A, Whitacre Mews, 26-34 Stannary Street, London SE11 4AB

British Library Cataloguing in Publication Information Available

Library of Congress Cataloging-in-Publication Data Available

LCCN 2017958894 | ISBN 978-1-4985-2666-1 (hardback : alk. paper) | ISBN 978-1-4985-2667-8 (ebook)

∞™ The paper used in this publication meets the minimum requirements of American National Standard for Information Sciences Permanence of Paper for Printed Library Materials, ANSI/NISO Z39.48-1992.

Printed in the United States of America

Contents

Acknowledgments

We would like to thank our fellow co-researchers of Nomadic University (uninomade.net) network, to whom this book is dedicated. We also thank Ana Luiza Lopes, who thoroughly helped the translation to English, for all the patient revisions and corrections through the many versions of the manuscript, as well as Daria Lavrennikova, who worked on the material as second translator. We are indebted to Raluca Soreanu and Thea Pitman, who have read the proofs and made very helpful comments.

Introduction

Accelerate the Biopolitical Process

THE OUTSIDE BEYOND DIALECTICAL TRANSITION OR MULTICULTURALIST TRANSLATION

In the 1990s, two theses were noted for diagnosing the new spirit of the time after the fall of the Berlin Wall. On the one hand, Francis Fukuyama foresaw the end of history, the end of the horizon of class struggle, at a moment when all horizons converged to the triumph of modern liberal democracies, within a globalized market economy, in which a breach becomes impossible.[1] The end of capitalism becomes unthinkable. On the other hand, Samuel Huntington announced the convulsive period of the clash of civilizations,[2] which would define the geopolitical lines after the disaster of real socialism.[3] The large gap between political-economic systems no longer exists; we are now dealing with blocks of values, identities, and religions competing with each other. For Fukuyama, in turn, there are no more nations in the antechamber of history, since we have already crossed the entire lounge, arriving at the white bedroom of post-history, where conflicts become local and territorial problems under the logic of global security management and market surveillance by the global institutions of financial capitalism. For Huntington, we still have a long way to go if the process of globalization is to be resolved in some kind of Pax Kapital, with geopolitical forces and whole armies entrenched around different civilizational axes, in an open conflict for hegemony that will define the future of Earth.

Both theses find correspondents on the left side of the political-ideological spectrum of representation. From the Left, Fukuyama's thesis of *the end of history* reappears in the disenchanted recognition of the triumph of neoliberalism, defined as a totality resulting from the last stage of capitalist globalization, after the post-Fordist transition (in the 1970s). In both the macro and micro-hegemonic version, however, neoliberalism tends to be founded under the condition of being a totality, governed by a structural-functionalist logic. The emerging processes that mix antagonism and exit always end up being sectioned and isolated in relation to the normal functioning of the system, often reduced to the condition of an explanatory appendix or occasional dysfunction. According to those analyses, going against neoliberalism entails an *exteriority* that is essentially and structurally disconnected from it. It follows that anti-neoliberalism ends up being reactive, merely resistant, transcendent, and implies the primacy of negativity. As for the hypothesis of the clash of civilizations, it reappears in the left-wing field in all sorts of neocolonial or postcolonial criticism that attribute to the underdeveloped, peripheral, or subaltern peoples the role of serving as a cultural, political, or anthropological *reserve* in relation to what would be the Western, capitalist, and modern matrix. The sometimes-dialectical contour reproduces old center/periphery dualist models of theories of imperialism, dependency, and uneven development. There would then be, only by their structural position in the global capitalist logic, "not yet" capitalist spaces, places of speech and/or preferential subjects for the practical formulation of an alternative to the capitalist civilizational horizon of the West—sometimes through a "politics of the excluded."[4] The culturalist praise of the non-Western or non-capitalist aspires to overturn Huntington, turning his thesis upside down, but its usual pretentious ignorance in relation to alterity, its sentimental and humanist sensibility full of multiculturalist and exoticized fantasies, all of this ends up confirming ethnocentrism which it was first intended to deny. The cleavage between the West and the Rest—another name for the Included and the Excluded—rapidly becomes a normative division, even if the signals are reversed. Instead of the old Enlightenment thaumaturgy, where modernity embodies universal reason, whose direction toward the future dispels the darkness of superstition, a new thaumaturgy is set to work: the anti-capitalist one of the oppressed but "not yet" fully subsumed peoples, who would come to save us from Ourselves and Western imperialism. To be outside of neoliberalism, in the latter case, is understood as the existence of other civilizational models or paradigms—often lined up as multicultural products (such as in the genre of world music), for the later consumption by the political sphere, often representative—an Outside that can be the "good living" (*vivir bien*) paradigm of the South American Andes, or the Ubuntu of Southern Africa, or the misty Chinese civilization, and its millenary teachings.

In short, the multicultural Western self re-enacts the intimate drama of self-alienation in relation to the Other/Outside, only to justify the actions with which it assimilates it in terms of its own problems, never in terms of the problems posed by the Other/Outside, for this would either be cultural appropriation, or inauthentic.

The first shift we propose in this book is in the very concept of neoliberalism. We understand that no matter how radical the rhetoric that diagnoses the laws, principles and theorems of the functioning of neoliberalism, it will always be a conservative approach of the state of things that the analysis will not be able to touch, even if stuttering and precariously, the emergent processes. These, before anything, give life and body to the organization of power and its networks in neoliberalism, and not some rationality of power that hovers over struggles, antagonisms and lines of flight. Therefore, instead of describing some macro-ideological or micrological totality, like a new Leviathan (although made of micropolitics), the intention is to explain neoliberalism as an ambivalent production of subjectivity, which is not resolved in dichotomies between included and excluded, interior and exterior, Capitalist and non-capitalist, Western and subaltern. Neoliberalism operates in a subjectivity of blurred lines, through flexible structures and amidst internal gradients and varying tensions. For this reason, we will return to Foucault's theoretical trajectory in *terra infidelium,* in his 1978–1979 lectures on the *Birth of Biopolitics*,[5] but to bring, to the nucleus of analysis, the positivity of biopolitical assemblages, flight, and biopolitical resistances. These are conditions of existence from which biopower is organized, a wrapping of technologies of power over the life of populations and bodies.

In the second case (the Outside is a non-place, a transcendent alterity), the attempts, generally on the left, were to rediscover an exteriority of capitalism or in other civilizations—the Chinese millenarian, the last indigenous autochthonous communities, the "anthropologist utopias"[6]—or in what would be an internal striation against neoliberalism, in favor of a more progressive or social regulation of the capitalist process—capitalism with Chinese characteristics, the progressive neo-developmentalist governments of Latin America, the geopolitical bloc of the BRICS. From this follows a kind of Cold War Ersatz, to console us of the end of the 20th century Cold War, when everything was so much easier. In certain cases, such as the identification of an Eastern rise of the "middle empire," pointing beyond capitalism, the formulation of this geo-theological Outside combines two different strands of anti-Western anthropology and anti-imperialist geopolitics into a mixed salad of appetizing transcendence for an authoritarian Left, the one who loves to speak about state, order, and regulation.

It is necessary to push even further the decolonization of Foucault's analysis because, in fact, the birth of biopolitics did not occur in Europe, nor was it first theorized by the ordoliberal and anarcho-liberal schools of thought, but rather in the colonial bio-enterprise in the South, as a particular configuration of the strategic relation with the Outside. From the outset this relationship was formulated by colonization, as a problem that appeared amid the organization of power and the governing of populations, material flows and races. Without exaggerating, the first theorists of neoliberalism were the Jesuits who faced the challenge of disciplining the savage body, paradoxically docile and indomitable at the same time. Neoliberalism tastes not only the saltiness of Portuguese caravels and the sweetness of the Colonial sugar mills, but also the marks of revolts and the blood of the slave and indigenous populations, whose uninterrupted and multidirectional mixture constituted the Body of the Poor throughout the last five centuries. And, from the beginning, it was constituted and traversed by the web of biopower. Thus, structurally, biopolitical society is both at the beginning and at the end of capitalism; it is at the dawn of mercantile capitalism informed by colonization, as well as in the last phase of late capitalism, on the threshold of its postmodernity, in the form of globalized post-Fordism.

Neoliberalism is nothing more than the globalization of the logic of biopower that erases the last frontiers and internalizes them in the control of actual bodies and its powers, in its affective and cognitive production, and in the machinic servitude that operates by signs, microtextures, and modes of feeling.[7] It is no wonder that today the planetary generalization of the precarious condition, fragmentary and full of informality, has been called *Planet of Slums*, the thirdworldization of the globe, and peripherization of the center, or even the Brazilianization of the world.[8] Once again, the risk in these formulations is to reduce biopower/neoliberalism to a totality, and thus to mutilate one side of the duplicity of poles implied in this relation of asymmetrical reciprocal presupposition. This leads to the drowning of the analysis, submerging the fact that capital must, as a *sine qua non* condition, establish a strategic relationship with the Outside, which is its own vitality. That is, instead of describing a neoliberal microtextural totalitarianism caused by its trans-frontier outpouring, the case is to observe how this fluid reality draws alongside it the conditions for an equally generalized subversion, a general tendency of multiplied lines of flight, of fractalized emergence, of equally polymorphous revolutionary becomings. The opposite of this approach would be to adopt a flattened understanding of globalization as pure negativity and totalitarian control, precisely fogging over the emerging processes of subjectivity that never ceased to be the real impetus of globalization itself.

FOLLOW THE MONEY

The third shift we propose is to follow the bodies to understand the relationship between money and capitalism in biopolitical terms. In *Living Currency*,[9] Pierre Klossowski focuses on the transition to post-modern capitalism, when image, mediatization and the "suggestion industry" become central to the process of valorization. Extremely dense, the essay published in 1970, together with erotic photographs and drawings, opened up space for a series of reflections on French post-structuralism throughout (Deleuze and Guattari, Foucault, Baudrillard, Lyotard). For Klossowski, the wage relationship that founds the social relation of capital should supposedly separate the direct access to bodies, to the extent that money is interposed between worker and employer. According to the liberal ideology, the direct exploitation of living bodies and their bundles of affections—fully possible in the previous regimes of slavery and servitude—would be overcome. The mediation of money, therefore, would be a barrier clause: placing bodily pleasures outside of commerce would give some protection to workers. In *Living Currency*, Klossowski shows, however, that money does nothing more than install a generalized, more intense circulation of one's own bodily affections. The bodies of the workers fill the free labor condition presupposed in the relation of capital, from end to end. The most perplexing moment of *Living Currency* occurs in the passage in which the author asks us to imagine an economy in which benefits are freely negotiated and then to say that this, in the end, already happens. Thus, if money should be mediation between productive bodies and capital, then bodies become mediations for money, the true governing force of production. Money ceased to be productive mediation: it is bodies that became productive mediation for money. Therefore, prostitution is not a specific activity, but the very essence of the capitalist mode of production, as Marx had already pointed out. In Klossowski, however, this does not take on an immediately moral connotation, as in Marx, inaugurating a whole tradition of moralistic criticism around selling the body, but it opens up the terrain of resistance for different strategies of perversion amongst capitalist relations. *Living Currency* points out that the transition to postindustrial capitalism and its financial dimensions brings the desiring body (the "voluptuous emotion") directly to the center of the capital process, when affective, aesthetic, relational, and imaginative aspects become useful for profit-making purposes and the circulation of money and values. It is the historical moment in which advertising, fashion, the world of celebrities, cinema, the entertainment industry, all this begins to act with primacy in capitalism.

What follows from Klossowski's analysis, then, is that the liberal theory of the wage earner already contains, within itself, the direct expropriation of the desiring body, which will fully come to the foreground in post-modern capitalism. That is, the turn to the post-modern does nothing more than realize the essence of capitalism: the conjugation between bodies and money, the logic of the living currency, the imperium of Eros. We will then use the insightful Klossowskian analysis to point out how political and economic liberalism, which accompanied the formation of industrial capitalism in the late 18th century, in principle already contains biopolitics. The trickery of liberalism consists of presenting generalized pimping as freedom of labor.

However, in capitalism's history, we have not seen any epochal turn to biopolitical production, to the aestheticization of value, or the *affective turn*. The mutation of money, increasingly more fluid and immeasurable, is what has historically determined the spread of direct corporeal investments in the process of capitalist subsumption of life. The functioning of the capital social relation, therefore, is structured much more by the dynamics of money than by the means of classical theory of value, based on a naturalistic metaphysics of labor and on the subsequent fetishism of commodity, which ends up placing full attention of criticism on the mechanism of alienation and ideology. The mystery of capitalism lies in money dynamics and not in commodity fetishism. For nothing is more densely puzzling, full of metaphysical subtleties and theological niceties, than money itself, which goes through successive transubstantiations throughout the historical contingent development of capital.[10] Klossowski was the one who paved the way to identify the path of the gradual coincidence between political economy and libidinal economy, which, two years later, will become one of the main theses of *Anti-Oedipus* (1972)—the theoretical landmark of Deleuze and Guattari's analysis of the current monetized regime of capitalism.[11]

In *The Birth of Biopolitics*,[12] Foucault identifies a strong difference between classical political economy, based mostly on the *negative* liberal principles of state containment, of the *laissez faire* and of the *ne pas trop gouverner*, and the contemporary neo-liberal economy, which assumes *positive* principles of intervention and constitution of civil society in the omnipresent form of the enterprise. However, we understand that this difference is not so influential and only serves as a distinction of degree and not of nature. Foucault himself, in the referenced course, notes how political economy inaugurates an art of governing art based on a practical reason quite different from the sovereignist political theology that is implicit in the *raison d'etat* of the old regimen. For us, the formation of neoliberalism does not change the sign of the principle of government, from negative to positive, but simply deepens this economic reason of government art to through and through governmentality. We could say that the economic theology of neoliberalism concludes the palatial coup that political economy applies on the political theology of sovereignty. But this happened well before the 20th century, a trend already present back to 1694, when the Bank of England, the mother of all central banks, started to replace the English Crown as guarantor of the circulating money. Therefore, from the beginning, capitalism already functions by thresholds and not limits, gradients and not abrupt transitions, and continuous cycles rather than teleological stages. In this sense, it would be necessary to reanalyze the classical political economists through the lens of the neoliberal theorists of the 20th century, and not the reverse, since it is the anatomy of Mises or Friedman that contains the key to the anatomy of Smith and Ricardo and not the other way.

But if biopower is at the beginning and at the end, within the mercantile colonial enterprise and within neoliberalism, what is it that has changed? From our point of view, the nature of money has changed, and it is this that can, retrospectively, best explain the unfolding of the mercantile, industrial and then fully post-industrial mode of capitalist social organization. If an internal acceleration of capital exists, as a permanent extension and conjuration of its operative threshold, it happens and must be explained by means of the successive deterritorialization of escapes, exoduses, miscegenation, and biopolitical resistances. And it is capital that pursues this biopolitical acceleration,[13] precisely, through the mutation of the currency and the always financialized dynamics of government, with which it seeks to ride these forces. The government of bodies and populations, in the America-Africa-Europe sixteenth-century trigonal circuit,[14] was the embryo, through a zigzag process of contingencies and mistakes, of the globalized financialization of capitalism in the 21st century. In other words, what we mean to say is that currency, central banks and stock exchanges were always the main government devices of capitalism, and not just after the transition to post-Fordism/postmodernism. The transition to the all-out financialization of life is not only due to the dismantling of the Bretton Woods normative framework, the oil shocks or the reaction to the global cycle of struggles around 1968, but it was the way capitalism could survive—enlarging and confining its limits—in the face of biopolitical resistances (of which 1968 is also a part of).

In view of this, it seems to be an entirely inadequate strategy today, under the all-encompassing almost redemptive label of anti-neoliberalism, to postulate regressive Outsides, in relation to biopolitical production at the basis of capitalism, be it in the form of a return to theory of value, to the state, to the Measure, to the Nation. For it is within neoliberalism (but before it), that emerging processes pulsate as intensive thresholds (an immanent Outside[15]), which continue to create tensions further beyond, by continually accelerating the process. Therefore, a *politics of bodies*[16]—of the desiring body, of the *savage body*, of the Body of the Poor—within the coordinates of the capitalist axiomatics must go through, as a condition of effectiveness, an articulation of a politics of monetary/desiring liberation, and thus liberate the living power of the desiring body that inhabits the relation of capital, disarticulating the relation with the Outside which capitalism depends on. More than a simple perversion, in Klossowskian terms, the case refers to devouring, anthropophagically, the very relation of capital.[17]

In addition to Klossowski, we rely on Jean-Joseph Goux's analysis of the shift of capitalism's imaginary, which emerges in the neoclassical theories of the marginalists, such as in literature, painting and the flourishing arts market, in the late 19th century. Thus, Goux positions the tipping point of postmodern economics, diverging from the usual periodization that places it in the wonder decade of the 1960s. Goux places it at least a century earlier. The marginalist revolution of León Walras, Jevons and Menger, in the decade of the 1870s, not only theoretically anticipates the formulations of philosophical and economic post-structuralism, but also points to a tendency already present since the formation of capitalism, which will later flow into neoliberalism tout court. Thereby the marginalist economists and modern artists of the period were the first to challenge the dogmatic metaphysics of productivists, utilitarians, and necessitarianists. Instead of relying on a metaphysics of production, the marginalists theorize that the dynamic nucleus of capitalism is, in fact, in the desire, in its unbridled, delusional creativity and its essential incommensurability. The notion of marginal utility professed by these economists allows one to understand that the production of subjectivity cannot be dissociated from the political economy, which should not be based on the scientificist premise of quantitative equivalence and objective value. It is necessary to relegate to a supporting role—since it is a later and imposed operation—the principle of equivalence implicit in the classical rationalist theory of value, the attempt of pinning down commensurability at capitalist production.

In this process, an economist like Leon Walras abandons the theory of value, the Archimedean point of analysis of capitalism for the productivists, in order to anticipate a schizophrenic theory of money that Marx had already outlined, in a glimpse, in the *Grundrisse*, and which would still await Bernard Schmitt and Anti-Oedipus in the second half of the 20th century, to be consistently re-elaborated.[18] For the marginalists of the 19th century, the model of the stock exchange, with its fluctuations, moods, and voluptuousness, is much more suited to the actual functioning of capitalism than the attempt to found it on productivist metaphysics, on calculations of the effective factors of production, or on objective time tables of labor division.[19] Upon resuming the marginalist revolution, Goux dismisses the flattened concept of money as a mere means of payment, the medium to facilitate exchanges, or the bulldog of commensurability between goods and wages. For Goux, money constitutes the principal instrument of governmentality, because it directly affects desire and operates in the two articulated regimes of capital—both in the molecular, by traversing the desiring body in the form of subjective debt economy; as well as on the molar, on the macroeconomic, banking and geopolitical structuring. To dominate the currency is the task that is imposed on a sovereignty displaced by the political/libidinal economy.

This implies that Goux's provocation, of which it is often better to revise the orthodox, neoliberal, or monetarist literature (as Foucault did in the referenced course), especially in matters such as finance, currency and credit, is more stimulating and penetrating than the parachute repetition of formulas and recipes by Left and Center-Left economists. The monetary factor is essential in capital, not simply to carry out the miracle of transubstantiation between bodies and banks, but due to its fundamental role in governmentality, biopower, and the continuous administration of the crisis intrinsic to the relation that capital establishes with its Outside. From this book, in very general lines of rewriting, we propose a history of capitalist formation which is a history of the transformations of currency, from its introduction as a credit-money by Italian banks at the dawn of mercantile capitalism and colonial expansion, to the disembodied predominance of fiat money and the full financialization of life in postmodernity, until it reaches our present day, with the delirious circulation of money as a pure sign, liberated from any static codification or fixed signifier, when the systolic expansions of growth prepare austere diastolic adjustments and vice versa, in a curious Nietzschean shuttle between Dionysus and Apollo. Once again, in order to avoid the risk of being paralyzed in the face of another despotic totality (even in the more postmodern shades of the molecular, of the biopolitical or of the "microtexture of daily life"), one must not lose sight of the existence of an intensive Outside, of emerging triggers and lines of flight, bubbling inside the living currency.

IT'S BIOPOWER, STUPID

Throughout the decade of 2000, a variety of governments in Latin America were occupied by Center-Left political forces. Known within the umbrella of the "cycle of progressive governments," in the beginning they divided themselves into two different rhetorical accents: on the one hand, more inflamed carnivorous and red versions that announced a new Bolivarian socialism for the 21st century (Andean countries: Bolivia and Ecuador, Venezuela, as well as Mesoamerican Nicaragua, with Cuba's endorsement), and on the other hand, more herbivorous and *rosé* versions, which openly assumed the synthesis between liberalism and social-democracy, a progressivism with a focus on the social (Brazil, Argentina, Uruguay, and to a lesser extent, Chile). But at the turn of the decade, despite the rhetoric, the whole cycle converged on the more syncretic matrix of the latter, inspired by the Brazilian government's arrangements with President Lula (2003–2010) of the Worker's Party. In the long run, the *rosé* aspect prevailed.[20] Developed from end to end by the theorists of Latin American progressivism, as a way to overcome the long neoliberal night of the 1990s, under the label of "post-neoliberalism," the so-called *lulization* of this Latin-American institutional cycle combined two driving principles: on the one hand, sustained increases in income through massive transfer policies, robust lines of popular credit and a wide range of social subsidies, and on the other hand, a strategic reorientation of industrial, tax and foreign trade policies, the attempt to frame a new matrix for national development within 21st century globalization, which in this book we have chosen to call, following part of the literature on the subject, *neo-developmentalism*.

It must be made clear that the conscious and deliberate conjunction between a new social focus and a new model of development did not really happen—not even in part—on the basis of the European Fordism-Keynesianism welfare or on the national-developmentalism of 20th-century Latin American theorists (in the manner of the ECLAC leading economists Raúl Prebisch or Celso Furtado), but rather on strictly neoliberal orthodox policies. The stabilization of currency in the 1990s, thanks to the consolidation of the macroeconomic neoliberal measures and the control of inflation, provided the structural conditions to build, in the next decade, the Lulist consensus, and then make the new social policy viable economically, politically, and monetarily. As a result, although the income-oriented policies of the period did not eliminate the extreme social/racial inequality prevailing in the continent, they were able to open up a small and unprecedented gap in the biopower bloc. Through this narrow breach, nevertheless, high-pressure flows of biopolitical productivity passed, resulting in a productive mobilization that injected huge amounts of energy into the social factories of living labor, increasingly more dynamical and fermented during the 2000s. With this, the deterritorialized currency ensured by neoliberal monetary policy was able to coalesce with the augmented social productivity of a newborn "new productive class," which already emerged within the flexible, mobile, and post-modern coordinates. The potency of the poor intermingled with the potency of deterritorialized money, bringing together desire and banks, labor and currency, economic dynamism and subjective debt-driven economy. The counter-effect of this post-modern flexible and financialized expansion of capitalism in the South was, as a line of flight, the generalized reappropriation of the instruments of capital. The counter-effect has become a counterpunch in capitalist expansion in the figure of a new *technologized body of the poor*, which eventually would turn against the historical violence of biopower. In this book this will be developed, since this meant the strengthening of a pole of autonomization and internal antagonism in the relation that capital establishes with the Outside, that is, it has led to the thickening of the threshold of the Outside that is intensive and subversive in relation to biopower.

To argue with the apologists of Lulism, who are confined to justifying governments from the statistical celebration of their policies, we called this mentioned line of flight from the Lulist period of transformations "savage Lulism."[21] By the "savage" quality of Lulism, we refer to the materialist philosophical concept of *clinamen*, which in the case means a deviation from the main tendency of the pink tide in Latin American politics. While the Brazilian government was boasting its "Brasil Maior" propaganda, we preferred to point to the subterranean affirmation of a *minor* dynamics, of an intensive class fracture within what was beginning to be called, rather apologetically and *sans rigueur*, the "new middle class." Until today, the mainstream discussion around Lulism orbits too much statistical-molar indicators, which are limited to explaining the processes of social inclusion, income distribution and market insertion of the bodies of the poor, but misses completely the point when we get to the biopolitical variables.

Lula was pleased to say that, during the period of his government, everyone could include meat on the menu. This immediately contrasted with the image of chicken, still the main source of protein among the poor in Brazil, and a symbol of the multiple recessions of the '90s, where even the middle classes were starting to stop buying red meat. In Lula's government decade (2003–2010), the shift in eating habits did not stop there, because the poor also began to eat the new productive and social relations that cooked and fermented throughout the first decade of the new century. New networks of communication and collective action emerged, new social tools, new arrangements and hybridizations of resistance practices. In social territories marked by a permanent reality of conflagration and enormous pressure exerted, above all, on the poor, the blacks and the indigenous population, in the city, in the countryside and in the forests, the new policies of progressive governments have unleashed a wealth *of* poverty. This rupture of dams brought the whole Baroque Latin American imagination of versatility, creativity and vital persistence to the fore, generating an unforeseen effect on a scale that surprised even the more optimistic policy makers. As a consequence, a major and profound transformation took place.

All those transformations were, however, very poorly metabolized theoretically and institutionally by the Left, in power (Workers' Party and allies) or in opposition (Trotskyists, more Marxist-Leninist Komintern-style and anarchists). From our point of view, the problem of Lulism was not an excessive complicity with neoliberalism, with merely short-sighted assistance programs or that would only correct conjunctural issues, as it was usually criticized by Leftists. In our view, in general, the problem of Lulism was then the timidity with which the progressive governments have interpreted the potential of opening a gap—*from within*—from the very flexible, fragmented and post-employment subjectivity that—as always had—characterizes the force of labor in the South, a force in permanent biopolitical acceleration. In our opinion, Lulism was not savage enough and remained too much state-centered, and when it was in flight and revolt (as was the case in the 2013 uprising), the Workers' Party's political forces turned against their bastard offspring.

But what, in the early years of Latin American left-leaning regimes, we might call political-institutional and theoretical shyness, gradually transformed into a frank reactivity. This occurred with special ferocity on the part of national-developmentalist and national-hegemonic sectors of the Latin American Left and broader Progressivism, who accused Lulism of being a project of the assimilation of the poor into a flexible capitalism, a malicious inclusion exclusively through the model of consumption and through the path of indebtedness, thus, leading to the reinforcement of informality and underemployment—to a re-edited dependentist thesis of "development of underdevelopment"—in addition to the old moral accusation of betrayal of the project of affirming a truly nationalist and anti-imperialist sovereignty. In 2017, after the real or imminent demise of these governments, the opening of the black boxes of Latin American neo-developmentalism shows us that this shyness, in fact, hid a political power entirely committed to the worst that there is in the biopower bloc in the South.[22]

The crisis of global capitalism triggered in 2007–2008 has led to a new inclination in the strategy of progressive governments. This inclination was not a proper turnaround. The newest internal realignment of the progressive cycle is better understood as a reinforcement of some preexisting tendencies, and the repression of others. The metabolization of the last major global crisis of capitalism by Latin American left-leaning regimes, in fact, led to the closure of ambiguities and diverse (often colliding) tendencies, the definition of an axis of government that would, henceforth, unify and conduct all the policies. In this correction of courses, the Brazilian case is exemplary in the trajectory of the government of the Workers' Party (PT) and Dilma Rousseff's governments (2011–2016), which ended up undermined by immense demonstrations, acute unpopularity and the approval of her impeachment by Congress, in August 2016. But in the beginning, around the turn of the decade of 2010, the Chinese demand had warmed up the commodity market to such an extent that the Latin American exporting countries enjoyed a corpulent trade balance, allowing considerable room for strategic decisions on social, infrastructural, and industry financing. At the same time, the growth of popular income, credit, and consumption has endogenously brewed the domestic market. The strengthening of the goods market catalyzed the national consumer goods industry, positioned between the influx of monetary resources from commodity exports and the cheapening of components coming from the giant manufacturing complexes of East Asia. Thus, in the face of the world capitalist crisis, without compromising the orthodox macroeconomic tripod (inflation control, floating exchange, and high interest rates), the Brazilian government adopted conventional countercyclical policies with a slightly Keynesian accent, by means of a robust package of investments and facilitations directed at the nucleus of export-related industries (agribusiness, heavy mining, the oil production chain) and the internal market (automobile, civil construction, home appliances), as well as encouraging the propensity to consume of the population in general. The success of the neokeynesian shield raised by the government economists against the crisis was great in the short term but misleading in long one. It has placed the country for a couple of years in the opposite direction of the recessionary global trend.

As a consequence, Brazil went through the turmoil of the world's stock exchanges almost unscathed, remaining with high levels of confidence in the rating agencies and, in 2010, reached a record mark of economic growth, the highest since 1974, with an increase of 7.5% GDP. With the "Chinese" performance of the economy and the positive perception of the effects of social policies, in the end of 2010 Lula's popularity reached its historical peak at 87% in the approval polls. It was his wonder year. It was not difficult for him to elect his successor to the party, a development-oriented technocrat, Dilma Rousseff, in the October 2010 election. Dilma, the first woman president of Brazil, a former guerrilla fighter, imprisoned and tortured by the military regime who, after the leaden years, made a career as a technical staff member in the mining and energy sector, personified the neo-developmentalist orientation in the decade of 2010, in its most authoritarian and economistic face. With her in charge, what was a tactical policy to confront the crisis and safeguard the complex arrangements at the base of social and income policies was consolidated in a perennial strategy aimed at framing a new model, a new longstanding project for the country.

Referred to pompously as the "New Economic Matrix" (NEM), the new political-institutional framing sought to achieve, so to speak, what would be "Phase 2 of Lulism"—that is, the moment in which an initial syncretism of neoliberal monetary bases and amorphous social and developmentalist policies would effectively give way to a state policy for full employment, substantial industrialization and technological catch-ups. In the theoretical and political-strategic formulations, the inspiration for NEM transpired with sympathy for China's long and steady developmental march, following the reforms of Deng Xiaoping, at the end of the 1970s, as well as a geopolitical reading of counter-hegemony against Washington-based geopolitics, centered on the BRICS and Beijing Consensus. These delineations reverberate deeper convergences and a curious parallelism between the developmentalist (from underdevelopment to development), socialist (from capitalism to communism) and capitalist (from feudalism to capitalism) transition schemes, all sharing a structural analogy around the same paradigm of the switch-over from backwardness to the modern, from the irrational to the rational, from the periphery to the center. It is enough for now to recall how the post-Maoist or "70% Maoist" synthesis led by Chairman Deng on the one hand re-emerged with financial globalization when it engulfed the American economy, in the middle of the post-Fordist restructuring of the 1970s, to such a degree that we can say that Chinese developmentalism is the other side—the rigid side of sovereignty—of the same coin, of neoliberalism *à americana*; on the other hand, the post-Maoist China resulted from the pacification of all the possible lines of flight of the period of agitation of the Cultural Revolution (1966–1976) and of the 1989 massacre, of the first massive occupy movement, whose epicenter was Tiananmen Square.

In Dilma's Brazil, as of 2011, instead of deepening income policies—for example, toward the maturation of the multiscale, successful and massified *Bolsa Família Program* as an institutional axis for universal income—the government has maintained and expanded the packages of aid for entrepreneurs, invested massively in the large agribusiness companies and practically assimilated the oligarchs of the civil construction in the tough nucleus of decision making, evaluation and planning of the actions of the government. Moreover, the Dilma government has made an incontestable and incontrovertible imperative from economic progress over what was considered "not yet developed," with no hesitation in stepping on minorities, *favelas*, and indigenous villages as if they were little flowers that, unfortunately, stood in the way of a country in the direction of its manifest destiny. A "national champions" belt of big companies was established around the government parties, in a virtual fusion between high corporate financing and political parties refinancing, through a flow chart of fees and deviations operated by banks and financial agents, both nationally and internationally. This sprawling scheme of corruption deepened and inscribed itself both in the neo-developmentalist model under the Workers' Party, swelling at the same pace as the GDP, so that the main Brazilian civil construction multinational, with business in more than eighty countries, maintained a "Sector of Structured Operations"[23] to share the billion dollar bonus of tips among political partners of almost every party. The molluscan corruption during the progressive government was indeed *structural*. In parallel, the optimistic tone of Brazilian progressivism not only rescued the dreams of greatness that rocked the country during the so-called "economic miracle," during the military developmentalist dictatorship (1964–1985), but also inspired the sovereign and anti-imperialist theorists who identified a chance to strengthen the national state within the game of geopolitical chess. This grandiloquence has been translated into the realization of international mega-events in the country: Military World Games (2011), United Nations Conference Rio + 20 (2012), Confederations Cup (2013), World Youth Day with the presence of the Pope (2013), the World Cup (2014) and World Olympic Games (2016), and in the aspiration for a permanent seat for Brazil on the UN Security Council. All these were perfect occasions and pretexts to further deepen the unashamed pillage of public funds, as well as reinforce a shock and hygienist urbanism in every metropolis, to remove tens of thousands of families from *favelas* and urban occupations, to enframe the popular economy and informal work, and to sweep the streets empty of the homeless and people in situations of chemical dependency or mental suffering.

The impossible neo-developmentalist model failed even on its own terms, as we lay out in this book, and finally collapsed in the middle of this decade.

A whole country attended the meeting marked with the future, but there it did not find the promised overcoming to the plight of underdevelopment. Rather it encountered biopower at its best shape. Phase 2 of Lulism did not lean toward the savage clinamen that we were theorizing at the time, but on the very logic of biopower which, in the first place, should have been contrasted: be it the neo-extractivist violence of agribusiness and dams construction (with emphasis on the pharaonic plant of Belo Monte, in the Amazon) against native people, small farmers, and the environment in general, or be it the equally neo-extractivist violence of the large constructions and the military pacification of the poor, against the inhabitants of the metropolis.

The monumental distance between the horizon of expectations and the actual experience broke out in June 2013 with the uprising of the multitude in more than 400 Brazilian cities, in a drastic sequence of direct actions, occupations of legislative houses and campaigns against institutional racism and the violence of progress (for example, the iconic "Where is Amarildo?" campaign), all of which inscribed Brazil in the global cycle of struggles initiated by the Arab revolutions just two years earlier. It was not by chance that the progressive government was efficient in coordinating repression in the diverse cities and to declare that, after June, peace should be absolute.[24] All of this was not a contingent or accessory diversion of the progressive government in Brazil, but a conscious and deliberate strategy that—even in the face of the violence and corruption that its policies clearly deepened—was carried out no matter what and to its ultimate consequences. Even doubled by the generalized population insurgency throughout 2013 (the all-time record of 2.050 strikes in one year[25])—which had also been one of the many expressions of a savage Lulism that turned entirely anti-Lula, anti-PT, and even understandably anti-Left—the shipwrecked neo-developmentalist matrix consensus of the (now over) progressive government continues to be defended to this day, in an unbelievable *narrative war* enveloped by sectors of the Latin American and also world-wide Left. In this critical oxymoronic support, justified by political (un)realism, some concessions to the criticism of the so-called "excesses" of those governments are often punctuated—always stating that this criticism, in fact, should not be done in public because "it is not the right time." Seldom has the nostalgic block of counter-hegemonism been so hysterical, rapidly clinging to the narrative plank of salvation which explains all political and economical collapse of the progressive government by means of conspiracy theories and a transcendental coupism, that would have struck them like a bolt out of the blue. The falling in love with power was really too extreme.

ANTHROPOPHAGY STRIKES BACK

If the first displacement was to understand neoliberalism as a globalized biopower and, therefore, as the peripherization and Brazilianization of the world, the second displacement, which we still propose,[26] consists in picking up the process to point out the emergencies, the becomings, that moment in which the periphery strikes back—that is, understanding neoliberal globalization in the late 20th and 21st century as the opening of a new and unavoidable terrain of tension, constituent of the production of subjectivities. The crucial clashes have been displaced to the plane of flexible and dispersive organization: "we want to think transcendence within the immanent, and it is from immanence that breach is expected,"[27] where everything is strategy, threshold, texturology, game of Chinese go. This means challenging the theoretical and political formulations, as well as their respective schematic and didactic obsessions, which insist on postulating an Outside beyond financialized and post-modern capitalism. This Outside could appear in the format of an idealized circuit of images in the form of spaces that are autonomous and purified of the capitalist logic, or else in the form of regressive utopias that adhere promptly to theoretical-political shortcuts: a return to the strong state, Full Employment and National Development, a reconstruction of Fordist Welfare (which, by the way, never existed in the South), or even a new Cold War, let alone a civilizing Outside, combining geopolitics and geohistory to identify the great enemy of "Western capitalism" in the Chinese "middle empire," which clearly sounds like a retreading of the culturalist thesis of Huntington. However, this does not mean to adhere to Fukuyama's thesis of the *end of history,* nor to any euphoria of the final dissolution as in the cosmogonies of absolute deterritorialization, of a certain accelerationism which flattens the tense ambivalence of biopolitical acceleration. We do not stop being confronted by the burning question: if we are always inside, what's left for the outside? In response, the second displacement does not imply liquidity as liquidation of the possibility of asserting another world—which would lead us to sink into the white dusk of postmodern irony—but rather a strong postmodernity, a liquidity that is strength and rigidity: an exit immanent to the regime of the deterritorialized money, an internal exteriority, an immanent outside. Given this, we can anticipate a conclusion: in the whole problematic of struggles and lines of flight, the identity of the Left has been functioning as pure transcendence.

Thus the strategic relevance to mobilize in the theoretical engineering that we do in this book, the concept of anthropophagy, as Oswald de Andrade has actualized it.[28] This effort stems from our uneasiness that, in the face of the political-economic collapse administered by capitalism, we must rid ourselves of the compulsion to the everlasting present and its consequent cynical political realism, at the risk of losing the ability to reconnect with the forces of the Outside, with becomings. The obsessive multiplication of exhortations to the struggle today seems to converge in the common grave of real paralysis, what unfolds in a maniacal exaltation of symbols that declare, de facto, the death of the diachronies. Regrettably, in the 21st century, the place of the anthropophagic manifesto of 1928 has already been homologized and secured in culture—be it national or world cultural history—as the original contribution of Brazilian modernism to world modernism. It occurs that anthropophagy does not return its mirror image to modernism, as if it were another product of the collection of icons to be cultivated (and commodified). What would mean reproducing to the maxim of the "first in Europe, then." In actuality, anthropophagy kidnaps and reinvents modernism, in order to give it back completely transformed, an alter-modernism. Therefore, we want to reactivate a strong anthropophagy, one that inverts the image that modernism creates of the South, so that it becomes possible to manufacture another image of modernity. Anthropophagy strikes back.

Firstly, anthropophagy cannot be reduced to a mere artistic-cultural intervention in a given political and cultural context, which would serve to establish a third way between cultural nationalism and the acritical import of canned goods manufactured in the first world. In fact, with Oswaldian anthropophagy, not only is the distinction between authentic national culture and imported colonized culture put into question, but also the notion of culture itself as the principle of differentiation between one and the other. In this sense, anthropophagy helps us to trace a line of flight in relation to the culturalist clippings traced, very often with extreme rapidity and convenience, between the poor South and the rich North, between the periphery and the center, or else, in so-called civilizatory coordinates, which are even more perverse, between East and West, between the West and China or India. The anthropophagic manifesto does not make a mere cultural criticism of the Brazilian situation as an underdeveloped country of not yet modern middle-class society. By linking a succession of delirious images with poetic intuitions and fragments of philosophical reflection, Oswald mobilizes a flow of fluent difference, a field of continuous deformation of entities[29] that he contrasts with the social fixed identities and psychoanalytic complexes of colonization and the colonized peoples. Throughout the torrential text of the manifesto, what occurs are various fulgurations: festivity, idleness, sex, nudity, cannibalism, primitivism, all sorts of tropical mystical visions, all of them resonantly mobilized together and metonymically against the quintessential figures of colonial civilization: the master of the plantation, the Jesuit missionary, the slave trader, the bureaucrat of the court, the provincial bachelor. From a Nietzschean point of view, Oswald thus summons his anthropophagic utopia against the sublimation of the psychic and rebellious energies that are now confined in the memory of the national formation, of the modernization of the barbarian lands. It is an anti-historical call and anti-memory conjuration to run against the grain of the psycho-social and historical-political syntheses of Brazil, like its theory of the three races as affluent "cordial" tributaries of the composition of the population. Anthropophagy allows the uncaptured fragments of insubordination, the barbaric threads of the national fabric, the minority becomings of races, their exodus, their deserts, all of this blasts its way through. In this sense, the anthropophagic madness is immediately historical-political and inaugurates a field of schizoanalysis of biopower in the South, which places the emerging processes—psychosocial and unconscious—in the foreground.

Instead of the introduction of a civilized form into a barbarous matter, an undifferentiated matter awaiting the redemptive arrival of Western-Christian civilization, we are dealing with an energetic declaration of love for the living and deformed matter of the primitive unconscious, a hylozoism of underdevelopment, a General Intellect of the decomposed and always deforming tropics. Primitivism, in this anthropophagic operation, implies the existence of an intensity of forces that go beyond the civilizational captures, a divine, wonderful and dangerous Outside,[30] a primitivism that is not at the origin, just as the Outside is not exterior, but in virtuality, in the intensive regime of the lines of flight, deterritorializations, and biopolitical wonders. Secondly, upon philosophizing on the practices of ritual warrior cannibalism of the Tupinambá, the Anthropophagic Manifesto points the displacement of culturalism toward naturalism reborn, toward the subject of the "natural man." It is not a soul that must be converted for Christ, for Progress or for the Proletarian Revolution. After an initial period of enthusiasm, at the end of the 16th century, the Portuguese Jesuits, educated in the School of Counter-Reformation, realized that the indigenous population did not raise obstacles to embrace deities, liturgies, sacred books and European beliefs. The indigenous population always said yes, without greater neuros, but there was a greater yes to the cultural-religious yes. That initially puzzled the missionaries. The question of the soul as a belief system mattered much less to them than the effective bodily, eating, relationship and warrior behaviors. Changing habits, *ethos*, ethograms, and daily pragmatics was the largest problem of catechesis, since, as Viveiros de Castro explains, upon reviewing Jesuit literature of the first centuries of colonization,[31] the body was primary—first ontologically speaking, and according to pragmatist ontology—in relation to the soul. That is why, in anthropophagy, one devours the relationship with the body of the other, the body understood as a bundle of affections, dispositions and capacities. Still, according to the anthropologist of the South American lowlands,[32] within anthropophagy what is in question is to elaborate an exchange of corporal perspectives, a zone of commutation and lateral cooptation between different perspectives. And the perspectives are always in the body, the body of everything that exists and produces effects in the world, in the same sense as in Leibniz philosophy where the souls are everywhere and are never scarce.

The result of the anthropophagic operation, therefore, will not be a spiritual conversion, it will not be to change the mind, to change opinions, to be manipulated or hegemonized by some ideology or culture, but a bodily and material metamorphosis, an increment of the *élan vital* that unfolds and immediately activates new practices and rituals of life. Therefore, when we return to anthropophagy, it is also a question of combating all the epistemologies of the South in vogue, which are allegedly supposed to add a new methodology based on the anthropological experience in the global South. This has nothing to do with anthropophagy, whose operators are bodies in continuous variation, an ontology of bodies and not an epistemology of souls, or in dialogue with Oswald: not an ontology but an *odontology*.[33] An europeanising epistemology is born from the narcissistic wounds of the self, since it is always an ego that, by assuming itself initially separated from nature, gives itself the task of rediscovering its relation with the exterior and restoring the lost primordial unity, thus reconciling oneself and the other, subject and object—ultimately in the self, in the subject, the alpha and omega of this metaphysical pilgrimage. This trajectory, metaphysically narrated in the Hegelian *Phenomenology of the Spirit*,[34] ends with Narcissus indulging in the honey of the very wounds he licks. This is a typically occidentalocentric problem, that of prospective solipsism, even when masked behind the most exotic multiculturalism, of an exotizing and complacent anthropological perspective, of an egocentric otherness. Anthropophagy does not assert that there initially the human and nature exist separately, but that the human is constitutively an integral part of the natural process, as well as the nature of the human, both created jointly in the process of production of *natura naturans*, of a productivist naturalism or a multinaturalism of the cultural (which is the reverse of the culturalist productivism of the natural, as in the implicit metaphysics of the law of value). And the point of departure will not be the solitary man facing the silent and seemingly unintelligible Creation. Rather it is the excess of nature, the anthropophagic superabundance full of dangers and wonders, and before which it is necessary to take on a perspective, to adopt precautions, to carefully bite the real and take an active role in a widespread predation of everything by everything, a generalized devouring chaosmos. There is Viveiros de Castro's concept of on(don)tological multinaturalism against all epistemic multiculturalism.

Oswald's primitivism is at the same time entirely postmodern. His new human being is the "Technologized Barbarian."[35] Not to see culture as a spiritual elevation in relation to nature, both higher and more constructivist and sophisticated in relation to a raw material, considered non-cultural and non-artistic in its essence, which always refers to the evolutionism of the primitive to the civilized, from the body to the soul, from the physical to the aesthetic. Its structural counterpart on the historical-political level disposes of the underdeveloped peoples in the antechamber of history, as primitives who are colonized, civilized, and capitalized for their own good, the Oswaldian figure of the Technologized Barbarian who places culture and nature in a coextensive field of perpetual interaction, one flowing over the other, a neo-primitivist society defined by its lines of flight. We were never catechized, says Oswald, just as we were never modern. With modernity turned upside down, the most postmodern is in the beginning, and the most primitive in the end. Or in dialogue with Viveiros de Castro: the post-mythic transformation falsifies the original mythopoesis. For, henceforth, we must speak of a techno-primitivism and not of regressive naturalistic preservationism, of a metabolic production of nature by nature, and not of some natural metaphysics that culture would process. The Technologized Barbarian updates, for the wonderful and divine superabundance of the South, Marx's concept of *Stoffwechsel* (metabolism): the simultaneous production of the social through natural mediations, and the natural through social mediations, encompassing the entire planet—that is, the Capitalocene. But if the line of flight is first, we prefer to say Schizocene. Following the indication of Beatriz Preciado,[36] thinking about capitalism in dialogue with Marx may even be enough when thinking from the waist up, but from the waist down, it isn't. This makes Marx and Marxism alone inadequate in thinking about an era of capitalism in which affection, pleasure, and fantasy are deeply embedded in the infrastructure of capitalism.

That is why Léon Walras, J. J. Goux, and Klossowski are relevant to our work, to allow for the necessary articulation between money and the desiring body, the principle tendency of the process of capital in the direction of identity, between political economy and libidinal economy, which only differ in the regime of functioning. The frequent spiritualization with which the Left pedagogically promotes the social struggles, to "change the views" of others, constantly seeing threats of capitalist manipulation everywhere, ends up denying the raw material of the body. And denying the body, it closes off the field of biopolitical strategy. As a matter of fact, this Occidentalocentric and Jesuitical Left had never been prepared to receive the anthropophagic vaccine. A whole leftist theme articulated as acculturation, "out of place ideas" and cultural imperialism needs to be replaced by the problematics of biopower, miscegenation, migrations, biopolitical resistances, the production of bodies and by bodies.

This is how we will discuss, at the end of the book, our last subject: the savage body. The *corps sauvage* is a new image of the body, rather than of the savage.[37] The savage body of the Technologized Barbarian is the Body without Organs, of the Global South. There cannot be a theory of subjectivity in underdevelopment without bringing to the fore the savage body. The body was often invoked theoretically to formulate organisms of power: the body of

Jesus or of the saints in Christian monotheism, the two bodies of the king at the beginning of sovereignty, the body of the tyrant (the face of Big Brother), the fearsome body of Leviathan, or even the body of criminal offense of criminal law.[38] The problem here is not so much the body, but its organization in the form of a power organism, whose head governs us from the heights of history, such as Hegel's Absolute Spirit or the Judeo-Christian God. The *savage* body channels the flows beyond the double wall—a market that imposes royal measures and a state that imposes an imperial citizenship—measure and affiliation, in fact, that are two plotted transcendences. We are not interested in a theory of national building, a theory of Brazil, guided by roots, "contributions" and identities, as a sub-question of Western modernity, the formation of national states, and the capitalist interstate system. Beyond the dependency, developmentalist, world-economy, or stageist modernizing theories in general, it is a case of abandoning their strange dualism, which always ends up referring to the theoretical-political analysis of evolutionism, to infernal vicious circles, and to the incessant reformulation of the old into the new, the archaic into the modern, the racist violence in the new markets and liberties. When we hear the importance of the word "progress" in the nationalist and modernizing discourses, and all its messianic declinations, it is biopower that, in the shadows, displays its toothy grin. It is the elites who are interested in what is theirs, because they constitute *their* national problem, *their* capitalist market and *their* sovereign state. We suspect that using Brazil or the South as an object of reflection is a thing of the place owners.

Just as we are not interested in peripheral modernism, since it is not a question of emancipating the underdeveloped through development (all of the mythology of the congested transition in the South), but rather to emancipate development itself by the underdeveloped (anthropophagy), becoming others in the process. The South's point of view of the North differs from the perspective of the North on the South, and this difference is first of all due to the fact that the South's perspective is located in bodies and not in ideologies, cultures, belief systems or epistemologies. The route from one to the other is different and asymmetrical. It is in this sense that we develop here what we have elaborated in the books *MundoBraz* (2009) and *KorpoBraz* (2014)[39] about the Brazilianization, the becoming-Brazil of the world, the activation of an altermodernity that never ceased to exist as a specter, haunting the process of colonial and capitalist expansion from its margins. A similar argument is put forth by Slavoj Žižek, who, quoting Peter Sloterdijk, writes: "capitalist globalization does not stand only for openness and conquest, but also for a self-enclosed cupola separating the Inside from the Outside."[40] We need not forget that this Outside is, under the circumstances, an Inside, a modulation of the Inside. The acceleration has always been in order to fracture modernity from within, in an alter-modernity, rather than trying to save it by means of a totalizing civilizing legacy, and do this without falling into the postulation of redemptive Outsides, mere shortcuts that soon turn into a boot stamping on our human faces. The way out is through the inside, plus intra!, biopolitical journey to the center of Earth, "to accelerate the process," in the Intensive Outside of biopolitical resistances, in the savage body, in the monstrous Body of the Poor exploited worldwide.

NOTES

1. Francis Fukuyama, *The End of History and the Last Man* (New York: Free Press, 1992).
2. Rodrigo Karmy Bolton, *Ensayos sobre razón imperial y mundo árabe contemporáneo* (Santiago de Chile: Lom Ediciones, 2016).
3. Samuel Huntington, *The Clash of Civilizations and the Remaking of World Order* (New York: Simon & Chuster, 1996).
4. See Kalyan Sanyal, *Rethinking Capitalist Development: Primitive Accumulation, Governmentality and Post-Colonial Capitalism* (New York: Routledge, 2014 [2007]). See also Slavoj Žižek, "How to begin from the beginning," in *The idea of communism*, by Costas Douzinas and Slavoj Žižek (London: Verso, 2010).
5. Foucault, *The Birth of Biopolitics: Lectures at the Collège de France, 1978–79*, trans. Graham Burshell (New York: Palgrave Macmillan, 2008 [2004]) as well as the immediately preceding course, *Security, Territory, Population: Lectures at the Collège de France, 1978–79* (New York: Palgrave Macmillan, 2007 [2004]).
6. According to Deleuze and Guattari's criticism of Pierre Clastres's anti-state anthropology, in *A Thousand Plateaus: Capitalism and Schizophrenia*, trans. Brian Massumi (London/ Minneapolis: Minnesota Press, 1987 [1980]), 358–61, 429–32. Basically, the authors question the postulation of a de facto (extensive and intensive) autarchy of the primitive societies against the state, which would be no more than an "ethnological dream," as well as the perspective that the insertion of the state into society operates, diachronically and from top to bottom, and not by way of internal gradients and thresholds from the beginning.
7. A delicate analysis of machine-like servitude (molecular/desiring regime) and how it is usually underestimated in relation to the logic of social subjection (molar regime), according to the intriguing book by Maurizio Lazzarato, *Signs and Machines: Capitalism and the Production of Subjectivity*, (London: Semiotext(e), 2014).
8. Ulrich Beck. *What is globalization?* (Cambridge: Polity Press, 1999).
9. Pierre Klossowski, *Living Currency* (London: Bloomsbury Academic, 2017 [1970]).
10. Thus, our analysis is a tributary of the Italian "operaist" (workerist) and post-operaist tradition, which in a revival of Marx will privilege a theory of money, very central to the *Grundrisse,* instead of a theory of fetishism, generally overvalued from a philological privilege focused on in the three first chapters of *Capital*. Accordingly, for everyone, Antonio Negri, *Marx Beyond Marx: Lessons on the Grundrisse* (Brooklyn, New York: Autonomedia, 1991 [1979]) and Christian Marazzi, "Money in the World of Crisis: The New Basis of Capitalist Power," in *Zerowork: Political Material* 2 (October 1976). Instead of pointing out an epochal turn from Fordism to post-Fordism after the Thirty Glorious, as the theorists of this field usually do, we prefer to follow the work of Jean-Joseph Goux (chapter 5) to rescue the reflections of marginalist economists in the 1870s (L. and A. Walras, Jevons, and Menger) as the first to displace a "Newtonian" conception of political economy, based on the theory of value and the metaphysics of production, toward a "Lagrangiana" or energetic conception from the desert of desire, its unrestrained flows, its immeasurable fulgurations. Marginalists have anticipated, in at least a century, the French post-structuralism and the schizophrenic theory of money flows by Bernard Schmitt. In accordance with, for example, Jean-Joseph Goux, *Frivolité de la valeur: Essa sur l'imaginaire du capitalisme* (Paris: Blusson, 2002).
11. Gilles Deleuze and Félix Guattari, *Anti-Oedipus: Capitalism and Schizophrenia* (Minneapolis: University of Minnesota Press, 1983 [1972]), 344, 345. As they put it, "Libidinal

economy is no less objective than political economy, and the political no less subjective than the libidinal, even though the two correspond to two modes of different investments of the same reality as social reality. There is an unconscious libidinal investment of desire that does not necessarily coincide with the preconscious investments of interest, and that explains how the latter can be perturbed and perverted in 'the most somber organization,' below all ideology."

12. Foucault, *The Birth of Biopolitics*.

13. Although there is a kind of moral panic when the word "accelerationism" is invoked, this seems to be a fundamental notion for us to understand the exodus of both living labor and biopolitical production in relation to the capitalist machine. In fact, the process of capital does not need to be accelerated, but velocity, of the conversion of a whirling flow into a laminar flow regime, from the continuous rectification of the clinamens according to the mononaturalist paradigms of transition or translation. Acceleration is the variation of variation and, in this sense, it captures the intensive and clinic(dynamic) dimension, which capital captures *a posteriori*, that is, *during a second structural moment*, although these forces of the Outside are attributed as if emanating from it as a quasi-cause. Therefore, capital only accelerates because it needs to pursue the crisis that, at any moment, threatens to escape its control, as a *fuite en avant*. The lexical puritanism around accelerationism is inscribed in an entire tradition reactive to postmodernisms and poststructuralisms, as "dangerous" theories or as simply accumulated to (in complicity with) neoliberalism. A good theory would be one without dangers . . . therapeutic. In the Baroque lands of the South, we could only warmly embrace the accelerationism, to problematize it from within. Such a morality, scandalized by its proximity to the enemy, as if threatening us with moral intoxication, is strange to us because it sounds profoundly anti-anthropophagic.

14. Luiz Felipe de Alencastro. *O trato dos viventes*. (São Paulo: Companhia das Letras, 2000).

15. "The outside is not a fixed limit, but a moving matter, animated by peristaltic movements, folds and foldings that together make up an inside: they are not something other than the outside, but precisely the inside of the outside." Gilles Deleuze, *Foucault* (London: A&C Black, 2006), 80.

16. As first proposed by *KorpoBraz: Por uma política dos corpos*, by Giuseppe Cocco (Rio de Janeiro: Mauad X, 2014), what we have developed in this book.

17. The concept of devouring, the enemy as immanence, according to the metaphysics of predation, of the anthropologist Eduardo Viveiros de Castro, "Immanence and Fear: Stranger-Events and Subjects in Amazonia," *The relative native* (London: Hau Books, 2015), 169–90. Anthropophagic predation is a *tertium gens* in relation to the transition x translation pair that mobilizes many debates about the coexistence of temporalities in globalization. If the paradigm of transition presupposes a teleological universalism and that of translation a multiculturalism, that of predation breaks with the mononaturalism implicit in both one and the other, to assert an ontological multinaturalism. See also Eduardo Viveiros de Castro, "Multinaturalism," in *Cannibal Metaphysics* (Minneapolis, Univocal Publishing, 2014 [2009]), 65–75.

18. Christian Kerslake, "Marxism and money in Deleuze and Guattari's *Capitalism and Schizophrenia*: On the Conflict between the Theories of Suzanne de Brunhoff and Bernard Schmitt," *Pahrresia*, no. 22 (2015), 38–78,http://www.parrhesiajournal.org/parrhesia22/parrhesia22_kerslake.pdf.

19. In turn, one may say that John Maynard Keynes was also a pioneer regarding this attempt to "rationalize the irrational," meaning to absorb, within parameters and analyses, something that was left outside the theory of value and its productivist approach. *Cf. The General Theory of Employment, Interest, and Money* (Macmillan and Co., 1964 [1936]). The Keynesian parametrization anticipates the synonymy of madness and capitalism that Deleuze and Guattari will develop in the 1970s, claiming that capitalism does not have a pathological mind, but is the very dementia of the *socius*. In any event, the authors recognize that "[o]ne of Keynes's contributions was the reintroduction of desire into the problem of money." *Cf.* Deleuze and Guattari, *Anti-Oedipus*, 230.

20. Pablo Stefanoni, "La lulización de la izquierda latinoamericana," *Le Monde Diplomatique*, September 2014, http://www.eldiplo.org/notas-web/la-lulizacion-de-la-izquierda-latinoamericana.

21. The concept emerged on the occasion of the effort to understand (and prolong lines of flight) the June 2013 uprising in Brazil, when we pointed out a resonance between the more intensive tendencies of Lulism and the popular revolt that spread through 400 cities of the country, contemporary and interconnected with the protests taking place in Turkey, around the social struggles of Gezi Park. *Cf.* Giuseppe Cocco and Bruno Cava, "Vogliamo tutto! Le giornate di giugno in Brasile: la costituzione selvaggia della moltitudine del lavoro metropolitano," EuroNomade, October 21, 2013, http://www.euronomade.info/?p=173. See also the dossier on the uprising of the Brazilian Multitude in 2013 edited by Giuseppe Cocco, "The Insurgent Multitude in Brazil," *The South Atlantic Quaterly* 113, no. 4 (Fall 2014).

22. For a synthesis on this debate see interview with Giuseppe Cocco in "Brésil, 13 ans après. Entretien avec Giuseppe Cocco," (Vacarme, June 5, 2016). http://www.vacarme.org/mot992.html.

23. Of course, we are talking about the mega-contractor Odebrecht, hand in glove with the Workers' Party in its multiple sub-imperialist enterprises across Latin America and Africa. When handing over the president's belt to Dilma, Lula assumed the role of main lobbyist of the company.

24. Paulo Arantes, "Depois de junho a paz será total," *O novo tempo do mundo—e outros estudos sobre a era da emergência* (São Paulo: Boitempo Editorial, 2014), 353–460.

25. "Greves em 2013 atingiram recorde e mobilizaram 2 milhões de trabalhadores," Rede Brasil Atual, December 22, 2015, http://www.redebrasilatual.com.br/trabalho/2015/12/greves-em-2013-atingiram-recorde-e-mobilizaram-2-milhoes-7006.html.

26. Following the trail which Giuseppe Cocco opened in *Mundobraz: o devir-Brasil do Mundo e o devir-Mundo do Brasil,*(Rio de Janeiro: Record, 2009).

27. Gilles Deleuze and Felix Guattari, *What is philosophy?* (New York: Columbia University Press: 1994 [1991]), 43.

28. Oswald de Andrade, "Manifesto Antropófago," *Revista de Antropofagia* 1, no. 1 (May 1928). For an English version of the text, see "Anthropophagic Manifesto," trans. Maria do Carmo Zanini, Sibila, March 24, 2009, accessed January 31, 2017, http://sibila.com.br/english/anthropophagic-manifesto/2686.

29. For a hylozoist discussion on the shapeless in the field of design, see Barbara Szaniecki, "Disforme Contemporâneo e Design Encarnado: Outros Monstros Possíveis," (São Paulo: AnnaBlume, 2014).

30. Like in *Divino, maravilhoso*, the viscerally anthropophagic Tropicalist song by Caetano Veloso (1968), especially in the ultra-deformed and delirious performance by Gal Costa (1969), https://www.youtube.com/watch?v=w7sbZkhdsFc.

31. Eduardo Viveiros de Castro, *A inconstância da alma selvagem* (São Paulo: Cosac Naify, 2014).

32. Viveiros de Castro, *Cannibal Metaphysics.* (Minneapolis: Univocal Publishing, 2014) [2009].

33. In a conversation with the Brazilian sociologist Antonio Candido, Oswald claimed that, "in times of universal devouring, the problematic is not ontological, but odontological." Miguel Conde, "Uma entrevista-aula com Antonio Candido na Flip 2011," *O Globo*, July 6, 2011, http://blogs.oglobo.globo.com/prosa/post/uma-entrevista-aula-com-antonio-candido-na-flip-2011-390689.html.

34. Friedrich G. Hegel, *Phenomenology of Spirit*, trans. A. V. Miller (Oxford: Clarendon Press, 1977 [1807]).

35. Andrade, Andrade, Oswald de. "Manifesto Antropófago," *Revista de Antropofagia* 1, no. 1[1928].

36. As Preciado puts it, "The real stake of capitalism today is the pharmacopornographic control of subjectivity, whose products are serotonin, techno-blood and blood products, testosterone, antacids, cortisone, techno-sperm, antibiotics, estradiol, techno-milk, alcohol and tobacco, morphine, insulin, cocaine, living human eggs, citrate of sildenafil (Viagra), and the entire material and virtual complex participating in the production of mental and psychosomatic states

of excitation, relaxation, and discharge, as well as those of omnipotence and total control. In these conditions, money itself becomes an abstract, signifying psychotropic substance. The Pharmacopornographic Era 39 Sex is the corollary of capitalism and war, the mirror of production. The dependent and sexual body and sex and all its semiotechnical derivations are henceforth the principal resource of post-Fordist capitalism." *Cf.* Beatriz Preciado, *Texto Junkie*, trans. Bruce Benderson from French (New York, Feminist Press, 2013 [2008]), 39–40.

37. Viveiros de Castro, *Cannibal Metaphysics.* Cit.

38. "The body provides an image for the much talked about 'body without organs', the great inspiration for Deleuze, who says that 'if we are to believe in the world, give me a body then'. The body isn't really the enemy, the organism is." Andrew Culp, *Dark Deleuze* (Minneapolis: University of Minnesota Press, 2016), 60–63.

39. The word "KorpoBraz" was first used in 1980 by the filmmaker and writer Glauber Rocha. See "Prefácio 80 (1980)," in *Revolução do Cinema Novo* (São Paulo: Cosac Naify, 2004).

40. Slavoj Žižek, *The Courage of Hopelessness.* Chronicles of a Year of Acting Dangerously (London: Allen Lane, 2017), 5.

Chapter One

Biopolitics and Development

Ten years later, the subprime crisis continues to be a historical watershed and "people abandoned the idea that 2008 represented a momentary blip."[1] The exhaustion of neoliberal globalization is marked by three great tendencies or inflections: the first is the scornful failure of the South American attempt to construct a neo-developmentalist exit; the second is the more and more unavoidable Chinese dynamics; the third is the combination of austerity policies and monetary emissions (*Quantitative Easing*) that caracterize the financial conduct of the Central European Bank (and of FED and of the Central Japanese Bank as well). The dramatic failure of the attempt to renew the traditional state interventionism in the sphere of Pink Tide in Latin American politics—in particular with the violent recession of the biggest economy on the Latin American continent, Brazil—shows and confirms that the escape from neoliberal regulation does not pass through the return of the traditional role of the state, nor through defining a "new" unknown role of the state. "(T)he Brazilian example" doesn't mean that "social abandonment nowadays has less to do with the willful exercise by governments of sovereign exclusion or direct violence than with their inability to subject the workings of transnational corporations to their own interests and those of their population."[2] The other side of the coin is that Latin American left-leaning regimes blocked the integration into the "global capitalism cupola" theorized by Sloterdijk and quoted by Žižek, and the consequence of this is greater and more violent exclusion and abandonment.

In this chapter, we go back to Michel Foucault's original interpretation of neoliberalism and to a reformulation of the enigmas of development in a biopolitical prospective. We propose a precise discussion of the experiences of the economic politics in the frame of Pink Tide governments in South America, particularly in the Brazilian case.

NEOLIBERALISM AS BIOPOWER

The Birth of Biopolitics[3] was one of the most controversial courses presented by Michel Foucault at the Collège de France. His public lessons in Paris, between late 1978 and early 1979, would only be published in 2004. The timing of the original lectures on neoliberalism was perfect. They were held a few months before Margaret Thatcher took office as prime minister of the United Kingdom (May 1979) and about a year prior to the election of Ronald Reagan to the presidency of the United States (1980). When Foucault presented his research on neoliberalism, the issue was still enjoying an air of novelty. "Neoliberalism" could be just another name for the great transformation happening in that decade, a set of displacement processes of capitalist formation, whose understanding would change depending on the approach the authors adopted to the *Zeitgeist*: entry to postmodernity,[4] change to post-Fordism,[5] or even the sociocultural genesis of a new "spirit of capitalism."[6] By the time Foucault's course was published, 25 years later, neoliberalism already constituted a key articulation piece at the public arena of debate, directly involved in the political disputes, and there was a vast amount of critical literature addressing it. At this point, denouncing the *pensée unique*[7] (one-dimensional thought) of neoliberalism had already been established as a watermark for the narratives and theoretical synthesis of the global Left, as a sign of belonging to the troupe. At the time of publication, the organization of the World Social Forum (WSF) counterpointed the Davos World Economic Forum (WEF), the summit of the financial capitalism main policy makers in the 21st century. Amidst accusations, neoliberals are thrown into an ideological hotchpotch, united at the service of the final dismantling of the welfare state of the Northern Fordism-Keynesianism. In the South, besides "natural" allies of governments involved with US imperialism, affiliated to the Washington Consensus, they would prevent the actual implementation of an industrialization regime which should eventually bring forth the welfare state. This worked as a kind of by-the-book anti-neoliberalism, which at the time had its moment of truth.

In this scenario, where neoliberalism appears as the main villain, the nuanced approach to "interdicted" theorists like Hayek, Milton Friedman, or Gary Becker held in *The Birth of Biopolitcs*; the precise distinctions the French philosopher makes between classical British and French liberals, German ordoliberals, and Austro-American anarcho-liberals, as well as some moments of resonance between the criticisms these groups make to state interventionism and to socialism and the genealogy of power so striking in Foucault's trajectory which could not go by without causing discomfort and even contempt in many readers who reached out to the course looking for yet more invectives. Such concern for nuances, typical of Foucault's work, was inadmissible when it came to theoretical contributions related to neoliberalism. *Delenda est* neoliberalism!

To these critics, the German *ordoliberals* or the neoliberals of the Chicago School could not be approached in many-sided terms, period. Even now, twelve years after the lectures of 1978–1979 were released (in 2004), collections of articles on *The Birth of Biopolitics*, the most commented upon of Foucault's free courses at Collège de France, are still forthcoming. Contemporary analyses oscillate between two poles: at one extreme, Foucault himself is accused of capitulating to the liberal ideology and, from there, his whole work would have definitely bowed to "post-modern" topics par excellence, such as the requirements for individual self-creation and authenticity according to a bourgeois "aesthetics of existence";[8] *Delenda est* Michel Foucault! Slavoj Žižek ironizes the French philosopher in these terms: "(His) notion of turning one's self itself into a work of art thus gets an unexpected confirmation: I buy my bodily fitness by visiting fitness clubs; I buy my spiritual enlightenment by enrolling in courses on transcendental meditations; I buy the satisfactory self-experience of myself as ecologically aware by purchasing only organic fruit, etc."[9] At the other, analysts identify a leitmotiv of renewal and self-criticism, with the purpose of an urgent redirection, essential in face of the epochal turn, on the eve of capitalist restructuring, and the inevitable conclusion that the Soviet state capitalism represented no alternative at all, that de-Stalinization would not save real Socialism.[10]

In this dispute, we are closer to the last pole. The thesis of Foucault's conversion to liberalism verges on the accusation of postmodern apostasy, according to the tradition of banning misguided and heterodox intellectuals. Behind the rejection lies the laziness and ineptitude to address and position problems considering their nuances and complexities. We cannot but agree with Lagasnerie when he says that the problem of neoliberalism works today as an eradication factor for theoretical and political divides.[11] The underlying issue consists of a containment effort that, identifying postmodernism as a dispersing threat, seeks to confine theories and practices of the Left—in other words, a clear imagination crisis of a disoriented and astonished anti-neoliberal Left, which often speaks the language of the state, order, and regulation. The image of a "neoliberal Foucault" is another scarecrow which was stuck in the theoretical and political generation of the 1968 cycle, whose "libertarian values" (this very idea, a tremendous reductionism) would have made the bed for the neoliberal turn. Foucault, as well as Deleuze, Derrida, Lyotard, among others, would be the fifth column of a more sophisticated Right, a postmodern Left questioning the *pures et dures* values of the state and sovereignty, the Party, the social welfare, and the class struggle. Even today there is still too much war over Foucault, because, to make matters worse, he remains a pop philosopher, with a legion of commentators, critics and enthusiasts. It has thus been hard to effectively stick any interdiction on him. We are "back to the future": the divide between Right and Left would be alive and determined by clearly distinct economic projects. On the one hand, a development project anchored in state intervention and the resuming of industrialization, and on the other, the neoliberal discourse at the service of banks' gains and international capital .[12]

THE POLITICAL ECONOMY OF BIOPOWER

The axis of *The Birth of Biopolitics* is the development of a new way of governing, new practices, verification criteria, and techniques with which the management of the living gains and maintains effectiveness. The birth of biopolitics corresponds to the gradual establishment of a "regime of truth," understood as the coupling between practices and discourses; not the adequacy of the idea to practice nor the logical consistency of coexisting propositions according to a rational system, but a problem related to efficiency, power, and the constitution of subjects within the network of powers. In a way, neoliberalism is a ramification of political liberalism, but the first completely displaces the latter, inverts it in different uses and innovates its working mechanisms. Foucault begins his enunciation describing the establishment of political economy as science at the dawn of the industrial revolution with the physiocrats, Adam Smith, Mill, and Ricardo. The new governmental practices interpreted by political economy have distanced themselves from the previous regime of truth, the reason of state (*raison d'état*) regime, characterized by mechanisms of economic mercantilism and the sovereignty principle in the political process of centralization of the European national states. If the reason of state is the art of governing proper to the general reorganization of power techniques and the constitution of subjects in the transition from feudalism to mercantile capitalism, political liberalism and the science of political economy are surface effects of a new reorganization, one associated with the development of industrial capitalism; the world market, whose epicenter was the 19th-century British Empire; and the many conflicts, tensions, and antagonisms typical of industrialization and industrial urbanization, their groupings and social segments. Neoliberalism will be, in this narrative of power and knowledge, the result of the next transition, from political liberalism to biopolitics tout court: the investment of populations' and individuals' whole lives, through new totalization and individualization processes, in the government matrix. Foucault will explain this in the course at hand through the socialization of the enterprise logic, the generalization of a new figure of subjectivity—the *homo oeconomicus*—and new knowledges related to the modulation of the self, whose first theoretical references were the Austrian romantic economic school, the German ordoliberalism and, later, the anarcho-capitalists of the Chicago School. In Foucault's terms, the consolidation of neoliberalism marks the final transition to a new way of regulating governmental practices in opposition to a government principle based on state reason, whether classical (mercantilism) or modernist (state capitalism or real socialism), but also diverging from ancient pillars of political liberalism and classic political economy. If, in the *raison d'etat* of sovereign societies, the government intervenes in the social and economic process from outside, like a *deus ex machina*—a force that arises marking a clear distinction in nature between the political process (state centered) and the economic one (society and market centered)—when it comes to the new power mechanisms of neoliberalism, a more internal regulation operates from within the social and economic phenomena, resulting in a less clear-cut contour between society/economy and state. The contour-lines of the state reason gradually become the tension-lines of governmentality. Instead of an external agent endowed with sovereign power and will, the decision-making core about regulation and its application dissipates in a cloud of small agents, which are themselves involved with the decisions that affect them, invested with an expanded economic rationality. So to speak, the biopolitics of neoliberalism operates rather as governance (Foucault prefers the term governmentality), diffuse and polycentric, than as a central government with a pyramidal structure of command. The market is the political-economic space where the neoliberal regulation operates and it coincides with civil society as a whole. In this sense, neoliberalism draws on political liberalism, but produces an infiltration of market rationality throughout the social issue, reaching the very production of subjectivities. One must only think that, according to political liberalism, the market already emerges as the "other" of the state, endowed with criteria intrinsic to the economic field and not to the political struggle per se. When it comes to contemporary neoliberalism a new logic of the market covers the entire field, leaving no space for autonomous political decisions, which would be not only harmful to the economy, but effectively impossible.[13]

The classical liberals claim that the action of the sovereign power must be circumscribed and limited, ensuring a domain relatively shielded from the harmful effects of governmental action. As they argue, the state and its centralized decision-making process also bring the institution of privileges, oligopolies, arbitrariness, administrative irrationality, favoritisms dictated by political conveniences, and so forth. The excessive intervention of the sovereign in the economic business means adding an element of unreason and disorder to the otherwise harmonious radiance of market rationality. According to classical political economy, governments should not interfere too much in the economic rationale, for the sake of the country *and the government itself.* When societies of the *Ancien Régime* and mercantilism are concerned, the assertion of central power is deemed necessary to dismantle the feudal and archaic elements and to open the possibility of a strategic plan on a national scale. In turn, when it comes to political liberalism, the economic dynamics has its own conducting logic, its own rules and internal optimizing factors, bringing about the principle of government self-limitation, of *laissez-fair* or *pas trop gouverner*.[14] These are the market mechanisms engendered in Europe at the turn of the 18th to the 19th century, theorized by classical political economists, which the neoliberal will adopt as a starting point for a new displacement.[15]

Foucault is guided by a genealogy that investigates the rise of new forces around the problem of power—not the ownership of power, but its exercise, its circulation, its subversion. Among all limitations to sovereignty, one the liberals hold most dear determines that the pricing dynamics should not be set from top down, organized by a power center with a clipboard filled with goals, production factors, and quota demands. Instead, prices must be gradually established, *pari passu*, through a polycentric set of actors, the iteration of supply operations, demand balances, risk taking, and technical innovations, which will eventually converge to a "natural price" given by the market based on the strength of the combined singular actions. The market becomes the regulatory principle for the geneses of prices and grants a "natural aspect" to the economic process, which would tend to be positive and self-improving.

According to Foucault, the new regime of truth of political liberalism is also linked to the European geopolitical balance in the 19th century, especially after the Napoleonic Wars. Unlike mercantilism, we are no longer dealing with fratricidal competition, often conducive to war, among the European national states, in a "zero-sum game."[16] With industrialization, the principle that someone must lose in order for someone to win becomes obsolete. The point in liberalism is the possibility of an endogenous development of industry and commerce which benefits, in a synergistic way, the totality of states encompassed by a continental market increasingly integrated. Liberalism thus implicates a new geopolitics, suitable to establishing an integrated market conducive to the emergence of new national industrial bourgeoisies, which would become the pillars of European power.[17] Still, according to Foucault, this endowed the old continent with an unprecedented coordination capacity among national economies, providing the institutional conditions for the beginning of an "unlimited economic progress,"[18] then spread throughout other areas of the globe—either colonies or satellite regions—in the form of the capitalist world market.

Foucault establishes a strong cleavage between classical liberalism and neoliberalism.[19] Even though one has the other as its antecedent, neoliberalism represents a fundamentally different regime of truth in terms of power mechanisms. Foucault contradicts *ante litteram* the whole critical tradition, which arose mostly in the 25-year gap between the lectures (1978–1979) and its publication (2004), that identifies the essence of neoliberalism with the topic of market deregulation, with the "belief in a mystical power" of economic rationality in itself, with the opposition to any vestige of state reason, and other clichés that testify more to an imagination crisis than it is an effective criticism of neoliberalism—which has been surviving, with just some hiccups here and there, during the various crises of the 1990s and the 2000s (Asian, Russian, dotcom, subprime, etc.). This is because, from liberalism to neoliberalism, and there is a displacement in the very nature of the exercise of power and its organization, there is a displacement in the art of governing with its complex bloc of devices, mechanisms, and discursive regimes, which Foucault synthesizes with the concept of "governmentality." Governmentality is something like an "assemblage," this concept that Žižek criticizes.[20] In other words, there is a change in how social and economic forces relate to each other: a different field of interactions between state and economy arises and a new kind of regulation falls directly upon productive activities. In this sense, neoliberalism is the regime of truth regarding a new functioning of capitalist accumulation, with a new approach to the production and circulation of value. Instead of platitudes about more state and less market, about minimum state or maximum market, Foucault prefers an analysis of how power works in the constitution of the "social," understood as the field where the subjects work in an intermingled way, from which something like a state or a market could be understood as surface discursive effects.

In reviewing the literature produced by the schools of German ordoliberalism and American anarcho-liberalism (represented by Eucken, Röpke, and Müller-Armack in the first case and Gary Becker, Milton Friedman, Stigler, and von Hayek in the latter),[21] Foucault examines how they understand the articulation among governmental practices and regimes of truth that define the neoliberal governmentality. Consequently, the question of governmentality does not focus on the simple definition of some historical totalizing structure, from some normative source, to explain the state power and market workings. Instead, it concerns the management techniques, methods, financial circuits, control devices, investment modes, as well as nucleations/distributions of calculation and decision power from which a decentralized government works, with what criteria and to what end. Neoliberalism is neither a revival of political liberalism, reissue of a commonplace version of Smith's "invisible hand," nor a mere sophistication of the full commodification of life ideology, nor even the opposition between Adam Smith's "invisible hand" and Giovanni Arrighi's visible hand of the state. If, in liberalism, the problem of power is guided by the need to impose limits to the state's reason so the autonomous economic sphere can provide the economic agents with enough freedom in order that the collective whole may function rationally, in neoliberalism, the problem becomes how the market economic freedom, rooted in the very social constitution, can determine the state as an internal principle of efficiency and optimization by the way of competition.

Instead of a local market under the supervision of the nation-state, responsible for the boundary conditions of a virtuous economic life, now is the national state which is under the supervision of the world market, where the last one watches the first by parameters of competitiveness, efficiency, and innovation.

The new phase of globalization is also anticipated by Michel Foucault a decade before the fall of the Berlin Wall. With the introduction of neoliberal governmentality, the economy ceases to be an empty market space, filled with commodity exchanges and individuals in a reciprocal rational action according to the autonomy of will and free enterprise, all through the state's discreet supervision; it then becomes an entire production of subjects as well as legal and economic mechanisms through which the civil society is created and governed. In neoliberalism, the market coincides with civil society, subsumed to the logic of capital. And society only exists within the diffuse field of the power of neoliberalism. The regulation occurs "from within," that is, at each point of civil societies' genesis and functioning, the space of modulated and flexible control, and not from external instances that apply governing principles or intervene when boundaries are disrespected. Thus, paradoxically, we start hearing more and more about the civil society—or the "third sector"—going into crisis at the moment it is subsumed under general productivity and governmentality controls. In fact, crisis becomes the operating mode of this new governmentality.[22]

In neoliberalism, the civil society rises as an autonomous existence with regard to the state reason, *because* it allows for a pervasive and endogenous organization of the social dynamics, whose primary aim is maximum efficiency and productivity. That does not mean it must be free from the hand of power, but that power operates in other terms, under widespread and polycentric conditions, as the cities becomes polycentric.[23] The nature of power exercise changes radically, even if those who formally own it remain the same. We are now talking about an eminently social and economic power. The change is qualitative in relation to the classical political economy, given a profound reorganization, described by Foucault as the historical creation of an "enterprise-society."[24] It is not about factories anymore, but companies; instead of workers, entrepreneurs. The propagation of investment and decision centers in the new neoliberal market occurs, not exactly by the proliferation of enterprises, but through a social corporatization, a pervasiveness of business practices in all activities and social fields. The labor composition is displaced: the employment logic, with its unitary and relatively stable wage relation, is replaced with the post-Fordist employability logic, in which salary operates in continuous variation, given by diverse projects and integrated from multiple flexible and transitory activities, until the wage relation is substituted by the relation between credit and debt. In neoliberalism, the very material life, the constitution of subjects, the meaning produced within discourses, all start to operate as a corporation according to the enterprise logic and, at the same time, the firms are diffused in the social and digital networks. In macroeconomic terms, the aggregated effects of this new subjectivity translate into a huge surplus extracted from productive flows running from the entire social tissue, and this now works much more efficiently than in previous governmentalities. Political liberalism is no longer adequate to grasp the new molecular institutions of neoliberalism, and classical political economy ends up losing sight of this immense flow of surplus, which goes beyond old formulas based on socially necessary labor time and other measures of value.

According to the Chicago School, and Gary Becker in particular, macroeconomics is too cumbersome to deal with the demands for efficiency and productivity that the social dynamics may provide, hence the need to deregulate it and directly invest at the microeconomic level. The idea is to ensure the optimal functioning of the enterprise-society against a disastrous interventionism and dirigisme of the state, when it seeks hegemony through inadequate practices and theories. Thus, it is no longer a simple matter of *laissez-faire* as in political liberalism, where the individuals compete for social goods in an inter-individual market, but a reconstruction of the *homo œconomicus* as an entrepreneur of himself (that is, the permanent self-modeling and training needed for employability), directly involved in a productive social fabric from end to end, which implies new practical and discursive subjective arrangements.[25] Far from the classical topic of liberalism, which opposes the individual freedom to the general will of the state, the political-economic workings of neoliberalism infiltrate into behaviors and regulate every aspect of life, including those not traditionally seen as economic or political, requiring a new science with no disciplinary boundaries.[26] Paradoxically, one of the effects is the becoming politics of all economic practices. At the same time, the economy invades all lived time, without rest, and without an outside.

In this new diagram of power, the social policies cease to be compensatory mechanisms for deregulations typical of the economic process and become direct devices with which the economy is formalized based on the enterprise logic. The main purpose of social policy is thus to reinstate the subject into the productivity and efficiency sphere, and keep him there at his maximum potential, so nothing is left outside the surplus game ruled by the renovated and all-pervasive world market.[27] This doesn't mean that politics and democracy are dispensable, but that market becomes the frame of democracy and politics the frame of the enterprise-society.

It is important to remember that neoliberals claimed the historical realities of Nazi Fascism and Stalinism as their main counterpoints. The horrors of German and Soviet regimes were used as a practical demonstration of the fact that the statist tendency represents a power organization which, reviving the *raison d'état* in the age of mass propaganda and mass production, becomes totalitarian and penetrates all spheres of life in order to fully dominate it. Foucault calls "state-phobia" this underlying mood of neoliberalism, which, in the paranoia display of the time, is a fear similar to the one caused by the atomic bomb.[28] While such state phobia, for good or for evil, has its ambiguities—it is worth mentioning, first, that Foucault's refusal of both realities is complete and, what's more, the diatribes of classical liberalism against the will of Leviathan were also ambiguous—Foucault points to an exacerbation of anti-statism among neoliberals, which leads them to extrapolate the rejection of the aforementioned regimes to the point of establishing a line of continuity from Nazism and Stalinism to the welfare state of Keynesian inspiration. This appears, according to Foucault, even in the criticism operated by neoliberalists against the American New Deal; the French *Front Populaire* (1936–1938); the Beveridge Plan (1942), in the United Kingdom at war; and, finally, the Marshall Plan, the recovering of Western Europe after 1947, under US leadership. According to neoliberals, all these political and economic recovery plans, related to the macroeconomic theories of John Maynard Keynes, carry the embryo of totalitarism.[29]

From Foucault's perspective, neither too much nor too little: we cannot attribute "to the state itself a process of becoming fascist which is actually exogenous and due much more to the state's reduction and dislocation," but neither can we put aside "the nature of the historical process which currently renders the state both so intolerable and so problematic."[30] As discussed, the case consists in tracing an analysis in terms of governmentality in order to grasp the exercise and circulation of power in their own right, instead of establishing some normative model of "good state" or "bad state," which quickly becomes a must-know guillotine at the will of those who apply it to reality. Conversely, there is also great inaccuracy, in a very similar fashion, to critiques directed *against* neoliberalism. Just as neoliberals flatten Fordism-Keynesianism into an embryonic state totalitarianism, there is an equivalent accusation, symmetrically inflated, which speaks of a totalitarianism of social relations after the neoliberal (or postmodern) turn, in the form of a "society of the spectacle" or a totalitarian consumerism. These criticisms also operate with all-comprehensive categories that end up invoking a power with capital letters, as if it was an inescapable totalizing machine to which any resistance would be useless and, at its limit, all subjective reaction would become nothing but an illusion created by the System on behalf of its perpetuation. For the latter, of course, it is imperative to conquer the state power, which is overvalued as condition enough to contain the totalizing advance of neoliberalism, a counter-logic that still operates in a world that does not exist anymore—that is, the world of political liberalism and its specific issues.

Foucault explains how neoliberalism is actually guided by a consistent rejection of massification, homogenization, and uniformization tendencies of the social organization based on mass industry and mass communication. Since German ordoliberalism, in the second post-war period, neoliberals theorize precisely *against* the mass society induced by Fordism-Keynesianism and the big industry disciplines, not the other way around. In this sense, some neoliberals responded better to the anti-disciplinary and anti-massification demands[31] from the '68 generation than a Left still attached to the totalizing and identitarian categories of the early 20th century.[32] Foucault shows how the persistence of anti-massification topics after 1968 provokes a transformation in capitalism, which, after the 1970s, began to restructure itself based on the diversity of social subjects, the market segmentation, the fragmentation of organizational bodies, and the mass Fordist working class dissolution.[33] The enterprise-society and the *homo oeconomicus*—hand in glove with neoliberal governmentality—are characterized, in fact, by greater social diversity and flexibility, by the dissemination of consumer and entrepreneurial micro-elements, seeking the overall and aggregate effect, which results in a much greater production of surplus.

> *Homo oeconomicus* is, if you like, the abstract, ideal, purely economic point that inhabits the dense, full, and complex reality of civil society. Or alternatively, civil society is the concrete ensemble within which these ideal points, economic men, must be placed so that they can be appropriately managed. So, *homo oeconomicus* and civil society belong to the same ensemble of the technologies of liberal governmentality.[34]

The fact that Foucault has named *The Birth of Biopolitics* a course focused on neoliberalism may seem strange to the reader, since the author has not addressed biopolitics in that book as emphatically as he had done elsewhere. Throughout the lectures, however, it becomes clear that a gradual formation of flexible technologies constitute the "social" advances toward a full investment of life "from within," as technologies internal to the constitution of the civil society and the *homo oeconomicus*. Classical liberalism devoted little attention to the problem of social work organization, limited to the time factor, the measure of value, and the allocation of factors of production (as in David Ricardo), thus producing a conveniently blind field concerning the social relations in themselves, which Marx extensively explored in his critical work on classical political economy. Neoliberals, on the contrary, not only claimed social work as *the* central concern, recognizing it as the locus of active economic force—decentering the factory unity and fostering "human capital," for instance—but also promoted the displacement of the political economy object by theorizing about the very constitution of subjects as efficient and competitive economical agents. Instead of an art of government—that, from a sovereign and unified center, plans the development and allocates productive factors on behalf of the national state—we are talking about a science that disseminates the regulation instances throughout the social fabric, instilling them into individual and collective subjects, favoring conditions where the economical agents themselves, in their social field of interactions, will conduct the economy in the most efficient way. It is a kind of decentralized interventionism, a spreading of specific and specialized interventions in multiple spheres of life, which, in the aggregated whole of the *enterprise-oriented* civil society, is conducive to greater productivity.[35] This implies a continuous (self-)modeling policy of productive relations, with final regulation coming "from within" the social, including that of the individual's relationship with himself. As Foucault puts it, "a policy of the economization of the entire social field, of an extension of the economy to the entire social field, but at the same time a policy which presents itself or seeks to be a kind of *Vitalpolitik*."[36] Thus, by replacing both state interventionism (of state reason) and the liberal principle of self-limitation, which frame the problematic of power boundaries between state and individual or politics and economy, neoliberalism emerges as "a positive program for the *laissez-faire*.[37] In conclusion, with the neoliberal governmentality, the market ceases to be guided by a self-limitation principle—as if the government wasn't supposed to be too much involved with the economy, unless when limit conditions or large-scale matters are at stake—and becomes an economic courtroom of permanent character, working from within the social field, without any moral or political boundaries, just tension thresholds. The point is no longer the mere individual subjugation, according to disciplinary power techniques or sovereign society technologies, through the elimination of criminal deviations and the suppression of unproductive dynamics. Instead, we are dealing with a positive regulation, applied on flexible lines, by self-transmutating modeling processes that don't emanate from a single center of power endowed with sovereign ownership (the state, the capital, the imperialism), as a pervasive modulation of the life environment—including the psychological one[38]—its conditions and its normal functioning. In other words, according to Foucault, the displacement from disciplinary technics toward the more modern biopolitics—the biopolitics *tout court*—is the same displacement that leads to the genesis of neoliberal governmentality, unfolding classical liberalism and, ultimately, totally displacing it, breaking from it.[39] The culmination of *The Birth of Biopolitcs*' lectures, when tracing the genealogy of civil society, is to demonstrate that biopolitics operates at the hard core of neoliberal governmentality—which, as Foucault indicates, did not go unnoticed by the most radical theorists of neoliberalism, who did not fail to theorize and systematize it, albeit in other terms.[40]

BIOPOLITICS OF DEVELOPMENT

Before, to quote himself in his conclusions of *Les aventures de la dialectique*, Maurice Merleau-Ponty wrote that it is always unbecoming to comment himself. But, he continues, "everybody who has published his opinions about vital problems is obliged to communique if he changes it."[41] So, let us go back to something of what one of us has written (in collaboration with Antonio Negri) at the beginning of the new century: even we don't change very much of this approach. The main task set by *Glob(AL): Biopoder e luta em uma América Latina globalizada*[42] was to reconfigure the political and theoretical debate on the problem of power in Latin America, based on the concept of biopolitics. It was an effort to rebuild the main discussions around uneven development or dependency theories throughout the 20th century. The great deal of effort was to displace the analyses and assume the tracks and becomings of *biopolitical resistances* as the main axis for a proper Southern turning point. The resistances in question include not only the struggles of slaves, indigenous people, immigrant workers and minorities, but also the biopolitical excess that escapes the diagrams of colonial and capitalist power: miscegenation, migrations, exoduses, and diasporas. It is something close (even if different) to what Jean and John L. Comaroff did about Africa, precisely when they write: "because it has plied its abrasive course in so many disparate contexts, at so many intersections of capitalist imperium, in other words, modernity has always been both one thing and many."[43] In this prospective, biopower forms a persistent diagrammatic bloc that runs through the various colonial and post-colonial periods addressed in the work. [44] Such bloc has not received priority of attention nor induced the approaches that were able to confront the problem of power within Latin America from the viewpoint of resistances. Specifically, we are talking about a chronic insufficiency in the Latin American debate arena concerning the theories aimed at explaining colonial and post-colonial realities, including national developmentalist schools whose cycle reached its peak in the 1950s; the dependency theories of the 1970s and 1980s; and the left-biased anti-neoliberal critique that emerged in the 1990s, culminating in a series of analyses that syncretize neoliberalism and the progressive governments under the label of "post-neoliberalism."

The approach in terms of biopower helps to relate the concept developed by Foucault in *The Birth of Biopolitics* to issues specific to Latin America, with which the concept ends up being displaced by the problematic field where it starts to operate.[45] The theater of forces with which biopower works, as it multiplies histories and subjectivities, has analytical primacy over the reductionism of a supposed progressive development of capital, which subdivides history in stages. We would then have the archaic and modern eras; or the colonial, monarchical (which, in the Brazilian case, included the 19th century enslaver empire), and republican periods; or pre-capitalism, then mercantile, industrial and post-industrial capitalism; or else a history that goes from underdevelopment toward development, from the periphery to the center of global capitalism; among other staged approaches. The idea is to admit historicism only to dismantle it. In which way did the birth of biopolitics, as theorized by Foucault, affect the development of capitalism and its economic theories? How did the transition from the *raison d'état* and the interventionist mercantilism to *laissez-faire* liberalism and, later, from the Keynesian revolution (1930–1970) to neoliberalism transform the development problem? We answer that, in the South, since colonization, the power *has always been biopolitical*. Our point here is not to refute Foucault's argument, but to tropicalize it by bringing his consistent analysis of power diagrams to the Southern perspective of the global history of capitalism. For us, the constitution of biopolitics—inextricable from modes of governance regarding bodies, populations, migration flows, and territories—is intertwined with colonization, and only later will integrate power technologies of modern Europe, until it transforms into contemporary neoliberal governance.[46] In his analytic of power, Foucault presents three diagrams: societies of sovereignty, societies of discipline, and societies of security or biopower.[47] This classification was never intended as an actual chronology or teleology which would go from sovereignty to biopower. Instead, this is better understood as a field of interactions where these diagrams relate to each other in a complex set of practices and discourses. Building on *Glob(AL)*, we argue there was a radical coexistence among the three diagrams during the colonial enterprise in the South, when biopower was born amidst the problems colonization had to deal with in the face of the biopolitical resistances, along both hard lines (direct struggles and slave insurrections, for instance) and flexible ones (miscegenation, migrations, and the like). The point is not treating the colonization history as precedent for capitalist development, as prehistory of its subsequent forms.[48] This would mean remaining attached to the universal history that capitalism establishes for itself as the horizon of intelligibility of the very historicity. Speaking of a "prehistory" of capital does not make sense unless we displace this idea from a period of time that would actually precede the proper capitalist social organizations to then approach it as a synchronic condition for the existence of capital processes, which must be continuously reproduced within fully established capitalist societies.[49] At the same time, this is not the case to produce an inversion between antecedent and consequent in order to make capitalism the expression derived from colonization—as if this change in order were the true founding metaphysics of modernity—which would mean making the same mistake and just changing the terms, but keeping the same concept of history. That is, the idea is not treating capital in any case as the source of history's intelligibility, according to an evolutionary line of progress, but assuming another concept of history, of temporality, one given by biopolitical resistances and the discontinuous and non-linear net they weave in the very flesh of colonization and capitalism. In other words, a perspective turning point that highlights antagonisms, discontinuities, becomings, and even defeats, and focuses on how they made an impact on the governmental practices of the living in a way that the biopower should be mainly understood as a continuous *reaction*, which involves a simultaneous and permanent work of contention and subsumption. The script for this reading against the grain goes through the colonial enterprise itself, the first industrialization of the 19th century, the modernizations of the 20th, and neoliberalism, as well as some theories involved in these transitions.[50]

In short, we reaffirm the new analysis method adopted in *Glob(AL)*, following two premises: (1) the history of capitalism as product of the very capital process, which presents itself as a linear progress from colonial pre-capitalist forms to advanced capitalism—that is, from the *raison d'état* mercantilism to neoliberalism, something that cannot be reduced to ideology, and is better understood as a "real abstraction" which produces effects and conditions the action of historical subjects; and (2) a second concept of history, discontinuous, with manifold temporalities, an intensive memory of struggles and exoduses, with a *biopolitical substance from the start*, taking ontological precedence over the reactive and derived forms that seek to control and exploit it. In dissonance with historicism, the book's conceptual and political bets, as well as its investigation of living forces that are able to foster a transforming project, are guided by something very close to Dipesh Chakrabarty's thesis of the two histories.[51] That is to say, it does not postulate that industrial capitalism, after emerging from the British Empire and then advancing under Western Europe's historical circumstances, had, from that point on, spread all around the world like oil that spills over a once clear gulf. In other words, without considering—as do some theories of uneven development and dependency—that the "miracle of capital" occurred first in modern Europe, then on the rest of the planet, in a uniform logic of dissemination. In this view, the local conditions outside Europe would be nothing more than objective obstacles with which capital comes across in its march—as if they were just external factors that, after enough time of breaking resistances and removing archaisms, would finally be subsumed.[52]

In an already classical cleavage, Chakrabarty calls "history 1" (*h1*) the line of progress of capital, defined by itself in retrospect as "universal history"—that is, the self-referential expansion of capital process toward the absolute world subsumption, which would bring the history of struggles to an end. The structure of *h1* remembers a Hegelian dialectics operating between the notions of archaic and modern, underdevelopment and development, where the dynamic subjective pole is capital itself, the actual subject of history. At the same time, the post-colonial Indian historian identifies a thriving parallel line, different in nature, but not exactly alternative,[53] that cannot be defined just as a non-capitalist space or, rather, a *not yet* capitulated or subsumed capitalist space "in the waiting room of history." For him, this other concept of history—or "history 2" (*h2*)—is not just the denial of capital, its absence, the non-capital, because it does not relate to capital only by opposition. It exists *in parallel*, in a complex relation of interaction and interpenetration, and continues to exist even after capitalist subsumption, according to different configurations. In fact, there is not only one *h2*, but a multiplicity of *h2s*, each one representing a different perspective "inside and against and beyond" *h1*. The two concepts of history—the (1) violently historicist, capital-centered, and stageist approach, which reproduces European archetypes of progress from archaic to modern ways of life, and the (2) living, non-linear, and perspectivist histories, which question the progress of progressivism—do not mirror one another as symmetrical poles, because they are asymmetrical and dialectically unassimilable by each other. That is why Chakrabarty prefers the term *translation* to the more usual *transition* when speaking of the transformations in the South. Transition is still confined to the progressive logic of evolving from archaic to modern ways of life, while translation seems more adequate to grasp a multidirectional logic where every *h2* is a viable point of view for another history—and each one can be subsequently translated into others. Through perspectivist translation, *h2s* can connect to one another as a plurality of temporal fragments, of "affective narratives," a specter lurking around capitalist "universal history."

On the one hand, Chakrabarty flanks the historicism implicated in the capitalist progress and its stageist versions—including the optimism of a civilizing mission inherent to capitalism—all of them following the "before in Europe, then in the world" sequence; on the other, he also circumvents an anti-capitalism defined by pure negativity—essentially a politics of the excluded or of the absolute Outside—as if there were something to preserve from capitalist impureness, some space and time not yet subsumed, in a kind of mystification of the pre-capitalist world. Instead of the "all or nothing" approach in the face of capitalist advance, the Hamletian time out of joint of *h2*s raises, in breadth and depth, the issue of strategy, the ability to trigger becomings that escape, flee, and overflow capitalist progress. The anti-modernity movements and discourses—and the *no* for capitalist subsumption they automatically express—thus embody a greater *yes* that determines their existence beyond any dialectics of progress or uneven development. In other words, they work as *biopolitical resistances*. It was in a similar affirmative sense, diagonal so to speak, that a "theory of subjectivity and underdevelopment" was addressed in *Glob(AL)*.[54] Thinking subjectivity within underdevelopment demands, first of all, considering the productive aspects of struggles, the dissemination of counter-powers and counter-perspectives (*h2s*) capable of formulating and asserting a project beyond capitalist biopower,[55] disrupting it. This must be done without mystifying an outside of power relations, as if biopower was an ideology that would be enough to denounce, and voilà, as if it was not an array of power practices, an entanglement of facts and knowledges, governmentalities.

By the end of the 1940s, assessments and theories on the development of economies thought as peripheral and dependent were gradually forming. These new theorization spaces emerged as a reflection of the moment when, in the face of the wars and depressions of the 20th century's first half, countries in Latin America, Africa, and Asia had to outline strategies to replace imports, promote industrialization, and integrate an internal market. Nourished by the European and North American political economy theories, and legitimized in the wave of the Keynesian revolution and the success of the Marshall Plan in Western Europe—which left a wide flank open for rethinking the economy beyond classical liberalism—"development economics" has come to address the particular and unfavorable conditions of the South. The main objective was to devise a consistent project of "autonomous" development that would not be at the mercy of world market fluctuations, foreign trade imbalances, or the benevolence of nation-states that completed their own transition to development in the 19th century. The point in common among national developmentalist theories in Latin America, especially in the 1950s, was the conceptual employment of a dualism between archaic and modern elements. It would be up to the development to overcome the archaic barriers in education, public policies, and labor organization, associated with old agrarian sectors and oligarchical structures, in order to enable the technical progress, the diversification of the "productive sector," and the qualification of the national proletariat. Another dualism present in these analyses is the dividing line between national capitals, whose decision-making centers would be concentrated within the country's borders, and foreign capitals, which would constitute a tendentiously plundering and alien force. Both dualisms—modern/productive *x* archaic/unproductive and national/sovereign *x* foreign/dependent—had major theoretical dominance until the mid-1960s in the strategies engineered by an arrangement between national-developmentalist think tanks and Latin American governments of the so-called "populist cycle" of the second post-war period. Thus, the alliances with the national productive business milieu were automatically considered preferable to unproductive capital (archaic, feudal, backward) or foreign financial capital (parasitic). The spirit of developmentalism involved in those dualisms implies esteem for the good and diligent employer at the same time it embodies the moral of the committed worker—both groups gathered to celebrate nationalism.

National developmentalist theories were also a strong trend within Left-wing parties. The Brazilian Communist Party (PCB), formally affiliated with the Soviet bloc, for example, adopted the thesis that the underdeveloped country should go first through a full bourgeois revolution. According to their perspective, this progressive move would precipitate a revolution of social forms, causing the spontaneous combustion of old oligarchies, large estates and other unproductive barriers. Only then could they proceed to a local version of the October Revolution. Without all that stageist process, there would be no appropriate human, technical, or organizational material that the Party could possibly organize to operate the socialist transition. Applying this logic, very similar to Moscow's Diamat, will enforce an alliance with national industrial bourgeoisie in the strategic plane. The good employers are deemed necessary in the sense they are considered a tactical ally for the cleaning of feudal and oligarchic waste, whose persistence would turn the actions for a social revolution unfeasible. Socialism is thus conditioned to the modernization and rationalization of production, mostly like the great Stakhanovite plans and, more generally, the mass conscription to labor in the Soviet Union.

Another current was created around the "ECLAC thinking," undoubtedly the most organic movement of political economic thinking that emerged in Latin America in the 20th century. The formulations of the Economic Commission for Latin America and the Caribbean opened a proper Latin American moment of reflection on uneven development, mainly after the inaugural work of the Argentinian economist Raúl Prebisch.[56] Diverging from the conservative theories of modernization, as well as the teleological dialectic of the Stalinists, Prebisch empirically identified a structural inequality in the balance of payments, in the relationship between industrialized "core economies" and "peripheral economies" specializing in primary commodity exports. That was due to several chronic mechanisms, such as a greater elasticity of demand when it comes to manufactured goods and a higher degree of organization among capitalists and workers from Europe, Canada, and the United States, who were better able to dispute among each other over profits and wages. The long-term effect was the deterioration of the terms of trade, so that the gains in productivity generated by modernization ended up absorbed to a greater extent by "core economies" in a systematic way. It directly challenged the neoclassical reformulation of David Ricardo's theory of comparative advantages, according to which each country should focus on its own geographical and historical endowments, since international trade would take care of distributing the productivity gains due to its own internal rationality.

For Prebisch, the structural mechanism of world commerce is intrinsically perverse because it makes it harder for peripheral economies to accumulate capital in order to invest in the transformation of their industrial park, leading to a vicious circle of underdevelopment. It is necessary, then, to devise a strategy that fosters the manufacturing industry under the unfavorable conditions in the South, proceeding to a controlled transition in a hostile international configuration. In other words, from a national developmentalist point of view, it is a matter of formulating an autonomous development path that cannot be a mere reproduction of the capitalist progress line as it occurred in Europe and in the United States in the 19th century. By the end of the 1940s, ECLAC thinkers theorized on a targeting strategy for investments and subsidies—in terms of changing the agenda of imports in a way that could favor an industrialization leap—known as import substitution industrialization policy (ISI).

During the 1950s, the ECLAC thinking evolved to focus not only on external factors related to international asymmetries, but on internal issues as well, always turning to the task of overcoming barriers and bottlenecks between underdevelopment and industrialization. According to the ECLAC Brazilian economist Celso Furtado,[57] technical progress would only come to a state that is capable of acting as a central economic planner. But such a state would not fall from the sky. It could only exist with a major transformation of the institutional structure, which meant eliminating archaic procedures, backward mentalities, short-term tendencies, and chronic low propensity to invest in productive activities and innovation. Moreover, the industrialization path should be linked to the development of an internal market, able to promote mass consumption, reduce underemployment, and distribute income. If the first formulations of ECLAC were focused on the structural inequality in the balance of payments in the terms of trade, they ended up covering a wider range of reform programs involving social inclusion through employment as well as the participation in the dividends of technical progress and market expansion. The strategic industrial policy was, thus, inseparable from a national project aimed at building a stronger and more structured middle-class society. In this context, an industry-based national capital would be once again the "natural" ally for a national developmentalist project able to reverse the pitfalls of unequal exchange between core and peripheral economies and take down the barriers of internal archaisms.[58]

In the late 1950s, when the import substitution policy gave signs of saturation, and industrialization alone had not contributed to the expected social inclusion and institutional democratization, national developmentalist economists, headed by Furtado, embraced a bold program of structural reforms, which were to be propelled by the growing social mobilizations. The "ECLAC thinking" seemed to have finally achieved a political subject. That was the maximum threshold that ECLAC dualists of the uneven development—directly involved with political forces of their time—stressed amid the bustling excitement of social movements in the early 1960s, soon after the Cuban revolution (1959). This threshold included the following program: *1)* strengthening of the national sovereignty in relation to foreign countries and the oligarchic-corporate resistances, with a focus on planned technical development and industrialization; *2)* interclass alliance between the most dynamic and productive sectors and the national bourgeoisie committed to the development; and *3)* democratic mobilization for carrying out the structural reforms. Three political limits derive from these programmatic spears that, *pour cause*, have assisted in the decline of this movement of thought: *1)* it did not question how to deal with other "backward" mechanisms and devices such as institutional racism, scientific positivism, and a certain attachment to authoritarian technocracy; *2)* it did not investigate who composed the "good employers" in their heterogeneity, as if there were structural reasons for a national industrial bourgeoisie to prefer an alliance with the working class rather than remain tied to oligarchic-corporate sectors or global capitalism; and *3)* it also did not research the struggles of its time, a lack of analysis of the class and social configuration of those who could constitute the driving force that was able to transform the terms of development.

The national developmentalist cycle in Brazil had a decisive setback on the occasion of the Brazilian military coup of 1964, which threw its main representatives into political exile and into a bitter theoretical swamp of skepticism. The reforms were conceived based on the existence of an enlightened vanguard to lead the process as well as on an Outside represented by the capitalist developed and/or socialist world. In this equation, the poor of the South appear as a mere object of the changes that should be brought or—less kindly, in the light of the Soviet Experience—imposed on them. It is true the autonomous developmentalist project of theorists that built on Furtado and Prebisch ideas[59] did not simply reproduce the models of uneven development imported from the mainstream political economy of the United States and Western Europe. However, this does not mean they managed to free themselves from "history 1." Even if they did not fall into a teleological stageism to the taste of the Soviet parties, ECLAC theorists aimed at the constitution of a socioeconomic situation in the image and likeness of the Fordist-Keynesian model operating in core economies. The problem was how to achieve, starting from underdevelopment conditions, such middle-class society with a strong welfare system, full employment, and mass consumption. But the "value of values" was not questioned enough. One could ask why it would be desirable to mirror in industrialized economies at the top of global capitalism (according to *h1*) in the first place. Besides, they did not take into account the workers' struggle perspective, with all its vital force, as in American Fordism, whose dynamism follows a strong class organization within unions, social movements, and local associations. This is not exactly a cultural or ideological problem, but one of political agency, organization, and production of subjectivity.

Of course, the situation analysis by Latin American national-developmentalists exceeded the modernization theories that chimerically prescribe, to the South, a similar path to that accomplished by the industrialization of the 19th century in Europe and the United States. Nonetheless, one of their central premises remains diagnosing an absence inscribed in the genes of the historical process, something that was never there (a defective *h1*), instead of pointing out particular genealogies and singularities capable of leading to different lines of rupture (*h2s*). This developmentalist theoretical approach proceeds through a negative perspective which is still anchored in *h1* and all its linear and homogenous timeline. The founding absence would come from the lack of either a sovereign fully formed state,[60] a robust proletarian trained class, or an integrated internal market. Thus, the question happens to be always enunciated in the negative: What have we not? In order to fill this absent structure, the authoritarian dimension of a usually voluntarist answer invariably shows up. Therefore, we can only detect a theory of subjectivity regarding the South condition in the classical literature of the Latin American national-developmentalist period through occasional flashes, glimpses, detours, and smoldering processes.

At the same time, when considered from the perspective of the terms of trade between established national states, globalization ends up being systematically criticized in a reactive way, as an everlasting threat, which triggers an "us-*vs*-them" posture, and as an ongoing structural barrier to the nationalization of decision-making centers committed to development. The underdeveloped countries would not have enough defense mechanisms to address foreign trade, foreign currency, and international division of labor disparities, which would render them an easy prey for the large multinational corporations, imperialist powers, and financial conglomerates of global capital. Once again, this somewhat flat view of globalization reinforced the (unsuccessful) bet in the national industrial bourgeoisie as the natural strategic ally for development. However, there is no good explanation as to why foreign capitalists would be more dangerously predatory than the ones at home. The foreign capital role is not discarded, but is considered, in itself, a barrier to obtain autonomous development—an idea of autonomy which assumes an international market populated of national states in hegemonic dispute, almost in a mercantilist fashion.[61]

According to Chakrabarty, "[t]his 'first in Europe, then elsewhere' structure of global historical time was historicist; different non-Western nationalisms would later produce local versions of the same narrative, replacing Europe by some locally constructed center."[62] What was really missing was building on another idea of "development,"—one different than the image of the capitalist future carved by the self-narrative of capitalism.[63] It was necessary to question more deeply not only the path of industrialization as a guiding principle to development,[64] but also the idea of a technical progress based on state planning, thus leaving aside all *biopolitical resistances* and the, so to speak, wild variables of the underdevelopment condition—all that which could not be reduced to an absence. These national developmentalist formulations, still marked by scientific positivism and an objectifying technical rationalism, only scratched "histories 2"—the specter that haunts power relations in the South and which can animate dependency and uneven development theories. Without disregarding the evident merits of the ECLAC school of thinking in relation to an affirmative Latin American movement of thought, their theorists were not able to create a theoretical position that could catch up with the virtualities and mobilization of *biopolitical resistances*.

FROM BIOPOWER TO A THEORY OF SUBJECTIVITY WITHIN UNDERDEVELOPMENT

In terms of power practices, biopolitics is at the intersection between neoliberalism and colonization, despite the differences in terms of spatial and temporal coordinates. Biopower is the grounds where antagonisms operate, where different strategies—reappropriations, subversions, reversals, transpositions, etc.—are at work in a myriad of friction zones, thresholds, and moving boundaries. The paradigmatic case of this operating method was the Portuguese colonization in Brazil and Africa from the 16th century on—the most radical experiment of colonial governmentality.[65] How could we not remember the poems of Fernando Pessoa in *Message* (whose title would initially be *Portugal*)? "And to the vast and possible ocean/Tell these escutcheons you see/That the bounded sea may be Greek or Roman:/ The sea without bounds is Portuguese."[66] The Portuguese imperial expansion beyond Europe was the utmost experiment of biopower, the molecular engendering and dispersion of practices that rule "from within," with an economy based on the flows of bodies, blood, land, races, as well as wealth displacements and modulations. To begin with, one cannot speak of Brazilian colonization without immediately referring to the colonization of the Western African coast, two processes not only simultaneous and closely articulated, but also coextensive in every aspect. Since the beginning, Portuguese colonization was built on black slave trafficking and Atlantic trade. This overseas endeavor began in the 15th century, establishing trading posts, warehouses, and fortresses along the West African coast. Then, by the end of that same century, they extended their enterprise to the Brazilian coast, just across the ocean. The Lusophone colonies on both sides of the South Atlantic thereby served as laboratory for the exercise of biopower and its art of governing.

Several centuries later, by the first decades of the 19th century, in the wake of the Haitian Revolution (1804), almost all Latin American countries were going straight from the colonial enterprise to being independent republics, unfurling national flags and carrying out their first nationalist modernization programs. That is, all Latin American countries—which were mostly Hispanic colonies—but Brazil. Such was the entrenchment of Portuguese colonial enterprise in the South Atlantic that, in Brazil, there was a "smooth transition"—or rather a non-transition—from colony to empire, a period that lasted from 1822 to 1889. That cry of national independency gave way to a slave empire that not only preserved the same families in charge of the agrarian oligarchic economy, but also extended the Portuguese dynasty, duly "Brazilianized," through the emperors Pedro I (1822–1831) and Pedro II (1831–1889). Consequently, the old slaveholder and baronial landowner logic was preserved, as well as the same governmentality on guard against rebellions, uprisings and riots, but also the more diffuse and subtle molecular resistances that erupted with an impressive frequency throughout the 19th century.[67] In here, after all, counterfactually, as the communist modernist Oswald de Andrade said, the *South won.*[68]

In *O trato dos viventes*,[69] Luiz Felipe de Alencastro describes how the Portuguese enterprise has expanded through a double articulated border across the South Atlantic routes, according to the transoceanic dynamics of the commercial triangle formed by Europe, the Americas, and Africa[70]—a continental border, associated with territorial expansion and domination, on both sides of the Atlantic. This continental border unfolds into a flexible, modulated blood border, which proliferates racial and social stratifications, producing multiple strata of mestizos.[71] The invention of the "mulato" through vertical miscegenation, for instance, enabled the colonial effort to forge a proletarian capable of carrying out a plural set of tasks, such as working at the farm, participating on occasional military expeditions to capture and eradicate troops of indigenous people, and even performing intermediary functions at the colonial chain of command, below the landlords or white freemen and above the slaves.

The Portuguese colonization already emerged, in the 15th century, as a deterritorialized organization, whose flows would either go through the smooth space of the sea or through the endless desert which the *bandeiras*[72] were obsessed with civilizing,—both places of trade, treasure hunt, body and soul conquest, and economic expansionism, but also of piracy, migrations, and all sorts of nomadism. According to Alencastro, the black slave traffic has given a strategic unity for an "Atlantic production mode," thus consolidating the centrality of slavery industry to the development of the Southern economic relations under the Portuguese shield.[73] The author did not fail to glimpse the structural relation between the power practices connected to the early capitalism in the Southern Atlantic and the formation, centuries later, of the classical political economy of the first industrial revolution in Europe. According to Alencastro, "liberalism does not represent a departure from the space of colonial policy, but precisely the opposite: liberalism is a reformulation of the colonial politics order without the necessary colonial ties."[74] The colonial enterprise cannot be understood merely as an incomplete capitalism, a rudimentary form, a "not yet," based on rapacious mercantilism and slavery. This actually corresponds to the biopower practices that, by conditioning and intermixing with the emerging capital process, will inform the political liberalism in the 19th century, establishing the "free market" and limiting the sovereign *raison d'état*. This is another case of what the post-colonialists call "periphery strikes back," meaning the power relations in the colonial periphery of European capitalism holds the key for understanding the capitalist process in the most developed countries. Following a comment of Karl Marx on E. G. Wakefield, we could go further and say that *the only way* to understand capitalism at the center is proceeding from its tension limits, from its boundaries. In other words, the only possible approach for comprehending it depends on an oblique perspective.[75] Nothing is missing in the colonization instances of power in regard to the "advanced" mercantilist nation-states of Europe. It is not that liberalism has overcome mercantilism and state reason, which would be the essence of the colonial enterprise as a kind of early mercantile capitalism. No. The very practices of the colonization process provided the deterritorialized government conditions, flexible and modulating, that were necessary for the genealogy of political liberalism. The economic synthesis of liberalism, based on market freedoms, could not exist if it did not embody biopower, which grants a degree of efficacy to domination and exploitation. This is first and foremost due to the biopolitical dimension of power, which conditions production control, circulation, and, ultimately, the "exclusion of the immanent resistance against the capitalist production born in the Atlantic."[76]

The *Pax Lusitana* in the Atlantic was a long-term project to pacify resistances, fixate exodus, and tame a fine mesh of refusal, sabotage and escape acts[77]—that is, a continuous recalibration of biopolitical control devices. Between Brazil, Africa and Europe, the *Pax Lusitana* had to deal with an endless source of seditions, rebellions, and uprisings throughout the whole colonization and subsequent imperial period. The extreme pictures of torture, killings, biological wars, and genocides are just the tip of the iceberg of a daily and normalized violence management without which it would be impossible to govern. The Brazilian sociologist Fernando Henrique Cardoso, when speaking about the myth of violence and crime, would say that the issue is not uncovering the "truth" behind the myth, but the *truth in the myth*—that is, "the structure of the myth":[78] how it works, what is its strength, its talisman. In a way, this talisman is a reproduction of slavery within the modernization stages, "the volcano where the society was grounded, which became the source of a situation of violence for both masters and slaves."[79] Not by chance, the volcano metaphor was used by the 19th century Brazilian novelist Jose de Alencar in his letters to Pedro II urging him *not* to abolish slavery: "Cut off this break (i.e., slavery), and a breeze will be enough to unleash a social war, the most bitter and hideous of all wars." He then asked, "Would you deem it a glory for your ruling, sir, to cast the empire on a volcano?"[80] The reproduction of the delayed abolition in Brazil can be identified in the absence of agrarian reform, in the nuances of racism, in the subordination of progress to order, all that which made the country and its cities "a volcano and a war more hideous for its senselessness."

We need to bring to the analysis both the diagram of biopower, which is consubstantial to the colonial enterprise, and the biopolitical resistances, which are internal to that governmentality. This means, for instance, identifying the elements of resistance also within the miscegenation lines, present since the colonization period. Back then, the colonial enterprise adopted explicit miscegenation strategies in order to enable the operation of several mechanisms of power and production. Those practices had such an extent that the entire history of Latin America remains unexplainable if the violent, subtle, and modulated racial governance they entail are not taken into account. Nevertheless, miscegenation is also a field of resistance, which present counter-strategies of "horizontal miscegenation" when those flows of bodies and bloods work against the grain. Migration is another example of biopolitical resistance, and that applies to both international migration processes—as in the case of Europeans, Asians, and Levantines that came to *fare l'America* in the context of the Long Depression in central economies, between 1873 and 1896, resulting in a massive influx of workers—and internal ones, according to lines that crisscross each other in superimposed temporalities: from the coast to the backlands, from east to west, from north to south, and from the countryside to the cities. The successive migration waves, encouraged by Latin American governments, caused a transformative fringe of urbanization and industrialization beyond any possibility of authority-guided planning, in an overflowing way. Not to mention those migration processes were one of the main factors for the organization of several groups—more specifically, the labor movement, radical unionists, communists, as well as the indigenous and the Latin American anarchist movements—in the early 20th century. The same can be argued for internal migration cases, such as the 17th century displacements of inhabitants from São Paulo that pushed agricultural, slave hunt, and mining borders during the colonial period; the complex settlement of a constellation of Jesuit missions and inland trading posts at large river banks (reaching deep Paraguay, Xingu and Amazon hydrographic basins); or the regular migrations from the Brazilian North-Northeast to the great centers of the South-Southeast, during all the 20th century, creating huge processes of self-organization and political agency from below, in order to build housing, infrastructure and services.

Instead of acting according to a disciplinary power of segregation (the apartheid type) and the molding of well-defined types, as in the United States or in Dutch South Africa, the Portuguese colonization worked, since the beginning, according to a continuous modulation of racial and blood flows.[81]

"The very Brazilian character called *bandeirante* is, in fact, the typical ideal of a dynamic that has the miscegenation as its central axis."[82] Instead of operating by rigidly stipulated and incompossible segments—the black, the indigenous, and the white peoples—a continuous miscegenation was endorsed and fostered by the very colonizing projects. This was an effort to expand the borders in extension (territory) and intensity (work force), which "has its roots in slavery, as well as in the thousand modulations of miscegenation, tracing back to the first actual import substitution policy—that is, to that moment when the *bandeirantes* from São Paulo captured 100,000 indigenous people to compensate for the slave trade interruption (1625–1650)."[83]

In a gray night when one could celebrate some friendly miscegenation of a properly Brazilian race—as some ideologists of racial democracy would like—the colonial biopower worked without eliminating differences. It rather strived to maintain the ideal types at the same time as they were distributed on a continuum, on a hierarchical gradation scale, functionally linked to the colonial government hierarchies. Instead of fostering distinct segregation, such continuous modulation not only promoted an orderly settlement of the land devastated by the invasion, but also served the efficiency of a social and political organization based on several "sub-races." The multiplication of sub-races, in the continuous line of miscegenation, promoted the circulation of power through the "law of gravity" of relative positions and scales, bringing cohesion to the whole. Violence can thus be employed by all and against all, in different degrees and modalities, according to the relative position of each one. This brought about, for instance, the "mulatto,"[84] another typically Brazilian invention in accordance to the racial engineering of colonization. The boundaries are blurred, they become gradients in order to be *reinforced*, so the normalizations and conditionings are diversified from within, as the immanent principle of the constitution of various identities and their relationships to one another—just as Foucault theorized for the moving and fluid networks of biopolitics in other circumstances.[85]

If biopower is the actual working ground of colonization practices and of Southern capitalist development, then this is the same ground where the resistance strategies—which will be, themselves, *biopolitical resistances*[86]—will operate as an antagonistic sticking point. Hence, besides that miscegenation as conformation strategy of the "Atlantic production mode," there is another miscegenation, a horizontal one, which stresses the molecular web of the (neo)slave state, producing tensions. The blood and territorial borderlines are intermingled with each other, as other kinds of lines, an embodied *nomos*.

Such possibility is inscribed in the very ambivalence of the problem of power, since the exercise of biopower presupposes the conditions in which it can emerge in another material sense, with a different effectiveness. It would be impossible for the biopower bloc to annul every migration and discipline every miscegenation process at the risk of disembodying the colonial enterprise, which needs the living force and *productivity* that the social relations of powers are endowed with. So, instead of simply suppressing migrations in a repressive way, it had to control them through a mosaic of more pervasive policies, such as public subsidy to prevent work-related flux, compulsory debt, sanitary inspection, and differentiated taxation.[87] The point is then identifying resistance processes from within the migratory flows, detecting proletarian responses to the attempts to anchor and discipline the biopolitical work force—in other words, the *biopolitical excess*.[88] Here again, we have the Foucaultian "birth of biopolitics" (which we call biopower in order to emphasize the possible antagonism), which is synchronic to the colonial enterprise, and so are the *biopolitical resistances*.

Alencastro speaks of the relative autonomisation of the triangular trade in the South Atlantic, between Brazil, Africa, and Portugal.[89] The historian delves into the singularity of this particular capitalist mode of production (also called "early capitalist Atlantic mode of production"), but his analysis does not mention any role for subjectivity production within the racist economy lying behind the slave regime.[90] However, in the longue durée, the deterritorialized smooth space of the Atlantic articulated with the striated space of the Brazilian backlands, which the Portuguese colonization and, later, the Brazilian neocolonial empire needed to keep territorializing through the *bandeiras*, supervised miscegenation processes, encouraged settlements, and migration control.[91] Until the first giant wave of migrations to Brazil came to an end, in 1930, the national unification project had to constantly negotiate possibilities between the indomitable externality of the sea-desert and the chimerical internalization of the backland-deserts.

The debate about Brazil (re)foundation in the passage from "Empire" to "Republic" is very important and actual. It has in its center the war of the new state against the community of Canudos.[92] The speeches of the rebel leader Antonio Conselheiro, founder of the messianic commune of Canudos (1893–1897), in Bahia, sound pregnant when he prophesizes that the backlands would turn into sea. He was talking about the exodus of the poor blessed leaving the landowners' yoke toward the free promised land.[93] Writer and journalist Euclides da Cunha vividly narrates the nomadic disposition of the Conselheiro's flock after winning what would be the first of many battles against the progressive forces then represented by the Brazilian army: "Having accomplished this feat, the pilgrims resumed their march, following the prophet on his hegira. They no longer sought out the towns, as before, but made the desert their destination."[94]

In an account that would become a Brazilianist classic, Cunha perceives, as if a sparkle, not only the biopower bloc befalling Canudos,[95] but also the underlying power of miscegenation in the blessed warriors: "the backlander is a fortress." It is interesting to mention that Conselheiro and the backlanders were not campaigning against the Brazilian Empire, which had ended in

1888, but the newborn Republic and the positivist modernization projects it entailed. The biopower bloc went unscathed by the change in government regime. The destruction of Canudos showed what really mattered was repeating and repositioning the same governmentality. In a similar way, Paul Gilroy, examining the Anglophone colonization in the Americas, identifies a drive of fantastic tones in the "Atlantic Ocean, bringing about more fluid and less fixed planetary cultures," in a book that highlights the diasporic power of the trans-African cultures of resistance, in their permanent mutation and spatial/temporal discontinuity.[96] Yet another example, *The Many-Headed Hydra*, narrates the uninterrupted transmission of exodus and resistance experiences in the Gulf Stream as a deterritorializing constituent power of a new class of rebels formed by levellers, diggers, sailors, adventurers, indigenous people, pirates, runaway slaves, and vagrants of all kinds, during the 17th and 18th centuries.[97] So, with Paul Gilroy we can take another ship: not the imperial British Leviathan as defined by Carl Schmitt, but a ship as a system of cultural, micro-political, and micro-political movements, producing intermediated concepts.[98]

The Outside of the nation-states in the South, until at least the mid-20th century, were these two figures of deterritorialization: the sea, from which could suddenly appear anything—a corsair ship, an armada, a monster—and the wild backlands, the rationality threshold beyond which one would find savages, El Dorados, miraculous fountains, and consciousness-expanding teas made of psychotropic plants. In short, the colonization encompassed two main lines of horizontal miscegenation, one from the sea and another from the backlands. There is still a schizoanalysis to be written, one that brings the unconscious to the colonization and vice-versa, to give voice to the civilization *delight*—but wouldn't it be the great task Glauber Rocha devised for his cinema?[99]

Even Foucault's lectures, with which we started the chapter, when they refer to neoliberalism entering the scene as biopolitical technology of power that unfolds political liberalism and breaks with it, seem—but only *seem*, as in *h2's* time is out of joint—to be late regarding the real functioning of Southern biopower. The attempts to explain the progress of capitalism, from liberalism to neoliberalism, succumb to the untimely and discontinuous history, which is far from linear and is even paradoxical, of biopolitical resistances.

NOTES

1. David Graeber, "Afterword—2014," *Debt: The First 5,000 Years*, (Brooklyn and London, Melville House, 2014), 393–97 [2011].

2. Jean and John L. Comaroff, *Theory from the South. Or, how Euro-America is evolving toward Africa*, (London: Paradigm, 2012), 185.

3. Michel Foucault, *The Birth of Biopolitics: Lectures at the Collège de France, 1978–79*, trans. Graham Burshell (New York: Palgrave Macmillan, 2008).
4. Jean-François Lyotard, *La condition postmoderne*, (Paris: Minuit, 1979).
5. Michael J. Piore and Charles F. Sabel, *The Second Industrial Divide* (New York: Basic Books, 1984).
6. Luc Boltanski and Ève Chiapello, *Le nouvel esprit du capitalisme* (Paris: Gallimard, 1999).
7. Ignacio Ramonet, "La pensée unique," *Le Monde Diplomatique*, January 1, 1995, see https://www.monde-diplomatique.fr/1995/01/RAMONET/6069.
8. Daniel Zamora, ed., *Critiquer Foucault: Les années 1980 et la tentation néolibérale* (Bruxelles: Aden, 2014). See also Zamora, "Can We Criticize Foucault?" Jacobin, October 12, 2014, https://www.jacobinmag.com/2014/12/foucault-interview. And also Mitchell Dean and Kaspar Villadsen, *State Phobia and Civil Society: The Political Legacy of Michel Foucault* (Redwood City: Stanford University Press, 2016).
9. Slavoj Žižek, *The Courage of Hopelessness*, 21.
10. Geoffroy de Lagasnerie, *La dernière leçon de Michel Foucault: Sur le néolibéralisme, la théorie et la politique* (Paris: Fayard, 2012).
11. Ibid.
12. Giuseppe Cocco, "Não Existe Amor No Brasil Maior," *Le Monde Diplomatique Brasil*, May 2013. This article anticipated in a few months the explosion of June 2013 Journeys in Brazil, when millions took to the streets imploding any binary division between Left and Right in terms of more or less State policy.
13. Foucault, *Birth of Biopolitics*, 18.
14. Those principles were first elaborated by the Marquis d'Argenson and René-Louis de Voyer de Paulmy in the 18th century.
15. Foucault, *Birth of Biopolitics*, 95.
16. Ibid., 53.
17. In Foucault's teachings, the genesis of political economy is not reduced to an ideological construct, as if the bourgeoisie sought to legitimize its hegemonic position in the transition from despotisms to the modern State or as if, when running the productive process, this same bourgeoisie produced discourses that would make it the general representative of the new regime. In fact, the knowledge that configures liberalism has its own materiality insofar as it is integrated to practices and discourses that condition the economic behaviors and actions of subjects. Hence the importance of tracing, not a history of ideas and ideologies in temporal succession, as a sequence of superstructures, but a genealogy of power in which the forces come into the scene and relate to each other directly involved in practices and discourses of a particular epistemic character or regime of truth. Michel Foucault, "Nietzsche, Genealogy, History," in *Language, Counter-memory, Practice: Selected Essays and Interviews*, ed. Donald F. Bouchard (Ithaca, NY: Cornell University Press, 1977).
18. Foucault, *Language, Counter-memory, Practice*, 54. However, according to Giovanni Arrighi, the heyday of liberalism in the 19th century corresponds to a prevalence of the free market ideology, a representation produced by the British Empire so its dominance can be presented as general interest for all capitalists. Circumscribed by this ideology, the free market prescriptions are understood as part of an overdetermination imposed by the hegemonic thought to an otherwise horizontal dynamic of economic exchanges. Arrighi essentially follows Braudel's scheme and Gramsci's influence, as well as the orthodox divide which opposes capital *x* market, where capitalism needs a hegemonic core in order to spread its accumulation cycles, from the Italian cities during the Renaissance to the American supremacy in the 20th century. *Cf.* Giovanni Arrighi, *Il lungo XX secolo: Denaro, potere e le origini del nostro tempo*, trans. Mauro Di Meglio (Milano: Il Saggiatore, 2014), 17–19, 175–92. In contrast, Foucault argues that the free market doctrine is not an ideology, but a discursive piece of a regime of truth, unintelligible outside the couplings with material rationalities and devices, which condition the working mechanisms not from above, by a hegemonic logic, but from a molecular web of capitalist power relations. There is no distinction of nature between the capitalist and the market spheres. It is as if Arrighi assumed Adam Smith's account as the explanation for the

real market mechanisms, rather than as a perspective on them—precisely the liberalist perspective. We go back to this topic in chapters 3 and 4.

19. Foucault, *Birth of Biopolitics*, 129–57.

20. Žižek talks about a "theory deployed by Latour and Delanda," but in fact he wants to criticize Deleuze and Guattari's very concept of "*agencement*." *The Courage of Hopelessness*, XVI.

21. Regarding Foucault's approach to governmentality and neoliberalism, we also rely heavily on Alexandre Fabiano Mendes, "Conflito e produção de subjetividade no capitalismo contemporâneo" (PhD diss., Universidade do Estado do Rio de Janeiro, 2012).

22. Michael Hardt and Antonio Negri, *Labor of Dionysus: A Critique of the State-form* (Minneapolis: University of Minnesota Press, 1994).

23. See Eduard Soja, *Postmodern Geographies*, (London-New York: Verso, 1989).

24. Michael Hardt and Antonio Negri, *Labor of Dionysus* Ibid., 175, 242.

25. Ibid., 215–25.

26. Ibid., 267–89.

27. Alexandre Mendes, Bruno Cava, and Giuseppe Cocco, "Foucault e os desafios para a América Latina," *Le Monde Diplomatique Brasil*, June 8, 2015, http://www.diplomatique.org.br/artigo.php?id=1897.

28. Foucault, *Birth of Biopolitics*, 76–77.

29. The problem with the concern of those neoliberals mentioned by Foucault is that they level three very different paths within the same Keynesian-type project of economic recovery which, in fact, converged, in the 1930s, with what Karl Polanyi called "dramatic resocialization of the market." The difference was that, in American, the New Deal did not lead to the pacification of the worker's struggles, providing grounds for conflict even if within the Fordist-Keynesian framework; while in Nazi Germany the relations with the working class were based on a tripod of extermination, cooptation, and consensus; and the Stalinist USSR was marked by widespread industrial discipline which impacted from the ranks of the Communist Party to the most everyday aspects of life, blocking any shred of labor democracy coming from below. Hence it follows that the living contradictions of the American welfare (New Deal) proved to be the most potent driving force among three paths with the same Keynesian background. For the complete argument, see Giuseppe Cocco, "Democracia e Socialismo na era da subsunção real: A construção do comum," in *O mundo real: Socialismo na era pós-neoliberal*, by Tarso Genro (Porto Alegre: L&PM, 2008), 55–91.

30. Foucault, *Birth of Biopolitics*, 191–92. Foucault's counterexample is the 3rd Reich. Instead of expanding some metaphysical principle in its own name, the state served as an instrument to the consolidation of the *Volk* (the organic community consisting of Germanic blood and people) under the loyalty and obedience regime known as *Fühertum* (principle of conduction). Thus, in the horrors of Nazi Germany, Foucault diagnoses more an inflated Party than the State presence, something he calls "party governmentality." *Cf.* Foucault, *Birth of Biopolitics*, 111.

31. Foucault mentions *The Society of the Spectacle*, by Guy Debord, and *Eros and Civilization*, by Herbert Marcuse, which passed through the hands of the '68 generation in the barricades of Western Europe.

32. Among those who also responded: from the struggles point of view, the Italian post-workerism in the 1970s, with Antonio Negri among others, as well as the sociologists of capitalism Boltanski and Chiapello. *Cf.* Boltanski and Chiapello, *Le nouvel esprit du capitalisme* (Paris: Gallimard, 1999).

33. It is also important not to make a mish-mash alternating massification criticism (Pedro Blas Gonzalez, *Ortega's the "Revolt of the Masses" and the Triumph of the New Man*, New York: Algora Pub., 2007) and social atomization criticism (Werner Sombart, *Economic Life in the Modern Age*, ed. Nico Stehr and Reiner Grundmann, New Brunswick: Transaction Publishers, 2001) as if they were (impossibly) simultaneous essences of capitalism. According to Foucault's method, the problem lies elsewhere and consists of scrutinizing the multiple individualization and totalization techniques with which the individuals and subjects constitute themselves and are constituted among each other in a coordinated manner, involved in the governmentality matrix. *Cf.* Michel Foucault, "A governamentalidade," in *Estratégia, poder-saber*, by

Michel Foucault, comp. Manoel Barros da Motta, 3rd ed., vol. IV, Ditos e Escritos (Rio de Janeiro: Forense Universitária, 2012), 281–305.

34. Foucault, *Birth of Biopolitics*, 296.

35. Ibid., 222–28. For a thorough discussion about the framework which leads to the entrepreneurial societies or "control societies," see Gilles Deleuze, "Postscript on the Societies of Control," *October* 59 (Winter 1992 [1990]): 3–7, http://www.jstor.org/stable/778828.

36. Foucault, *Birth of Biopolitics*, 242.

37. Henry Calvert Simons, *A Positive Program for Laissez Faire: Some Proposals for a Liberal Economic Policy* (Chicago: University of Chicago Press, 1949). *Apud* Foucault, *Birth of Biopolitics*, 216.

38. Foucault, *Birth of Biopolitics*, 259.

39. In an article that refers to Foucault's lectures addressing the effects of neoliberalism, economist Luiz Gonzaga Belluzo makes a mess. The confusion already starts in the second sentence, as he qualifies neoliberalism as "ideology." It gets worse when the author tries to amend the Collège de France Professor for not bringing the "predominance of commercial forces" to the foreground. Belluzzo completely distances himself from Foucault when he claims that the main effect of neoliberal governmentality consists of the commodification of social relations, exactly the opposite of what Foucault believes (Foucault, *Birth of Biopolitics*, 147–49). Belluzzo then circumvents the analysis of the entrepreneurial society as a molecular web which increases productivity and competitiveness, to speak of the concentrationary trend of multinationals and, as a consequence, the dispute for hegemony between China and the United States as the "new world order." The point is that none of these topics have anything to do with Foucault's approach to neoliberalism. As small and prosaic as Belluzo's article is, it serves as evidence to point out two phenomena: 1) the role neoliberalism plays as Judas doll, an anti-panacea whose assumptions and theories, immediately reduced to whatever clichés, does not deserve a detailed analysis, and 2) the ease with which neoliberalism functions as synecdoche for capitalism, which often helps to exempt those who are affiliated to the so-called "post-neoliberalism" current, as if it was "less capitalist," "social capitalism," or even "anti-capitalist" in some cases. Luiz Gonzaga Belluzzo, "Foucault e o neoliberalismo," *Carta Capital*, May 24, 2013, accessed February 2016, http://www.cartacapital.com.br/revista/750/foucault-e-o-neoliberalismo-1420.html.

40. No wonder Foucault pays special attention to Gary Becker, from the Chicago School, whose focus on the "social" covers topics as diverse as gender, race, migration, demography, crime, and education. According to Melinda Cooper, Becker was a pioneer in answering the demands of the anti-normalizing struggles of 1968, anticipating the capitalist restructuring of the 1970s and 1980s. Melinda's criticism is related to the fact that, in the face of the general objection to gender roles and family normativity of the *soixante-huitards*, Becker prescribes the loosening of the marriage contract; the relativization of the idea of "deviance"; and the introduction of the business logic, focused on productivity and profitability, in the family context. The case here is not so much a family business, as old as capitalism itself, but the corporatization of family relations. According to her, the notable focus Foucault dedicates in his lectures of 1978–79 to Becker, perhaps the most radical among the American neoliberal economists, is symptomatic of a period when the French philosopher was sympathetic to the neo-fundamentalism of the Iranian Revolution of May 1979, which later, as we know, would have a tragic outcome for gays, alcoholics, and women. *Cf.* Melinda Cooper, "The Law of the Household: Foucault, Neoliberalism, and the Iranian Revolution," in *The Government of Life: Foucault, Biopolitics, and Neoliberalism*, ed. Vanessa Lemm and Miguel Vatter (New York: Fordham University Press, 2014), 29–58.

41. Maurice Merleau-Ponty, *Les aventures de la dialectique*. (Paris: Gallimard-Folio, 1955), 315.

42. Antonio Negri and Giuseppe Cocco, *Glob(AL): Biopoder e luta em uma América Latina globalizada*, trans. Eliana Aguiar (Rio De Janeiro: Record, 2005).

43. *Theory from the South*. Or, how Euro-America is evolving toward Africa, (London: Paradigm, 2012).

44. In its negative sense, the "biopolitical bloc is thus based on this state of exception and is always a choice that only the racist modulations are capable of making tolerable for the elites.

The existence of two prison systems (formal denial that citizens are equal before the law) can only be explained in these terms." Negri and Cocco, *Glob(AL)*, 112.

45. When referring to the South—in this book topically situated in Brazil, Argentina, and Mexico, we do not make a definite distinction between first and third world, North and South, rather indicating the existence of an internal threshold, a grey border area of variable size, which folds and refolds the North and South opposition in multiple layers. There is also a North of the South and a South of the North, as well as *n* intermediate variations not readily classifiable according to one or another extreme.

46. According to Deleuze, the disciplinary societies act by molding, like a chisel hitting the marble, while the societies of control operate through continuous modulation, like radio frequencies and their wavelike nature.

47. Michel Foucault, *Discipline and Punish: The Birth of the Prison* (New York: Vintage Books, 1995 [1975]). See also Foucault, *Security, Territory, Population: Lectures at the Collège de France, 1977–1978* (Basingstoke: Palgrave Macmillan, 2007). It is worth mentioning that what Foucault describes as societies of security or biopower correspond to what Deleuze calls, *mutatis mutandis*, societies of control.

48. See Karl Polany, for instance, for whom liberalism could not be explained without reference to merchant capitalism in the Atlantic and slave labor. Karl Polanyi, *The Great Transformation: The Political and Economic Origins of Our Time*, 2nd ed. (Boston, MA: Beacon Press, 2001 [1944]). See also Otávio Velho: "it is an indisputable fact, although increasingly recognized, and here we go back to a more global perspective, that in addition to the huge profits and capitals generated by the slave trade and slave plantations, the dynamism they imprinted on the world trade between the sixteenth and nineteenth centuries, and the new structures of capital, labor, production, and service they mobilized were fundamental and indispensable to the development of European capitalism." *Cf.* Otávio Velho, "A ideologia da miscigenação e as relações interraciais no Brasil," *Revista Lugar Comum*, no. 29 (February 2012): 117–27, accessed March 2016, http://uninomade.net/wp-content/files_mf/110610120215A ideologia da miscigenação e as relações interraciais no Brasil - Otávio Velho.pdf.

49. Jason Read. *The micro-politics of capital: Marx and the prehistory of the present* (Albany: State University of New York Press, 2003).

50. On this subject, our references are Yann Moulier Boutang's *De l'esclavage au salariat: Économie historique du salariat bridé* (Paris: PUF, 1998); Dale Tomich's *Through the prism of slavery: Labor, capital, and world economy* (Lanham: Rowman & Littlefield, 2004); and especially the writings of Sidney Mintz on American plantations understood as "capitalism before capitalism." *Cf.* Sidney W. Mintz, *Sweetness and Power: The Place of Sugar in Modern History* (New York: Penguin, 1985).

51. Dipesh Chakrabarty, *Provincializing Europe: Postcolonial Thought and Historical Difference* (Princeton: Princeton University Press, 2000), 47–71.

52. We are referring to the uneven development theories in general, which, one way or another, consider the historicism according to the logic of the industrial capital advancement around the world, since its constitution in England by the end of the 18th century. This is a diachronic scheme between before and after the capitalist subsumption, a dualism between center and periphery based on a linear and homogeneous time.

53. Chakrabarty avoids any alternativism, as if there was an Outside of capitalism that deserved safeguarding in its own right. According to him, the relations between the history of capital and the living history that escapes it can be of many kinds: opposition, overlapping, negotiation, miscegenation, etc., according to different strategies. Chakrabarty, *Provincializing Europe*, 66.

54. Negri and Cocco, *Glob(AL)*, 41–51.

55. After all, "history 1" is a retrospective narrative in which capitalism institutes itself as universal history, thus suppressing the struggles and movements that had to be faced along the way. Chakrabarty, *Provincializing Europe*, 98–100.

56. For the founding text of ECLAC structuralism, see Raúl Prebisch, *El desarrollo económico de la América Latina y algunos de sus principales problemas* (Santiago: CEPAL,

1949). From the same author, see also *Problemas teóricos y prácticos del crecimiento económico*, (Santiago: CEPAL, 1973 [1951]).

57. Celso Furtado, *A economia Brasileira: Contribuição à análise do seu desenvolvimento* (Rio de Janeiro: A Noite, 1954). See aslo Furtado, *Uma economia dependente* (Rio de Janeiro: A Noite, 1956). For additional contact with the work of the Brazilian economist over the years, see his compiled articles in *Essencial Celso Furtado* (São Paulo: Companhia das Letras, 2013).

58. Negri and Cocco, *Glob(AL)*, 63–68.

59. Ricardo Bielschowsky, *Pensamento econômico brasileiro: O ciclo ideológicodo desenvolvimentismo* (Rio de Janeiro: Contraponto, 1995).

60. In the South, we would lack a molar, unified, and strong state, which would be capable of determining economic dynamics and escaping the harmful influences of globalization. This myth, according to which the power relations in the South would be "weaker," would justify even more biopower in its most brutal version. The absence of a state compatible with European sovereignty and *raison d'état* does not mean that the molecular web of biopowers—regulating life "from within," from the bodies as well as blood and racial flows, in a continuous modulation—is less violent, less effective, or less racist (on the contrary).

61. What José Luiz Fiori calls "inter-state capitalist system." *Cf.* José Luis Fiori, "O Sistema interestatal capitalista no início do século XXI," in *O mito do colapso do poder americano*, ed. José Luis Fiori, Carlos Medeiros, and Franklin Serrano (Rio De Janeiro: Record, 2009).

62. Chakrabarty, *Provincializing Europe*, 7.

63. As Marx, in a teleological passage, wrote in the Preface to the Book I of *Capital*: a "country that is more developed industrially only shows, to the less developed, the image of its own future." Karl Marx, "Preface to the First Edition," in *Capital: A Critique of Political Economy*, by Karl Marx, vol. 1 (Hardmondworth, 1990), 91.

64. "Thus, the sequential or staged character of the process is responsible not only for the ease with which it can be brought underway, but also for the lack of training in technological innovation and for the resistances to both backward linkage investments and to exporting that are being encountered. The most important consequence of sequentiality, however, is the fact that it has become possible for industrialization to penetrate into Latin America and elsewhere among the late latecomers without requiring the fundamental social and political changes which it wrought among the pioneer industrial countries and also among the earlier group of latecomers." Albert Hirschman, "The Political Economy of Import-Substituting Industrialization in Latin America," *The Quarterly Journal of Economics* 82, no. 1 (1968): 31.

65. Negri and Cocco, *Glob(AL)*, 77–81.

66. Fernando Pessoa, "Mensagem," MENSAGEM Ilustrada E Comentada, accessed April 2016, http://www.inverso.pt/Mensagem/MarPortugues/Padrao.htm. Poem translated by João Manuel Mimoso based on Mike Harland. And there it is his poem "First": Europe lies, reclining upon her elbows:/From East to West she stretches, staring,/And romantic tresses fall over/ Greek eyes, reminding./The left elbow is stepped back;/The other laid out at an angle./The first says Italy where it leans;/This one England where, set afar,/The hand holds the resting face./ Enigmatic and fateful she stares/Out West, to the future of the past./The staring face is Portugal. Pessoa, "First," MENSAGEM Ilustrada e Comentada. Accessed April 2016. http://www.inverso.pt/Mensagem/Brazao/castellos.htm. Also translated by João Manuel Mimoso based on Mike Harland.

67. For instance, Monica Dantas, comp., *Revoltas, motins, revoluções: Homens livres pobres e libertos no Brasil do século XIX*. (São Paulo: Alameda, 2011). See also Eugene Genovese, *From Rebellion to Revolution: Afro-American Slave Revolts in the Making of the Modern World*. (Baton Rouge: Louisiana State University Press, 1980).

68. Oswald De Andrade, "Aqui foi o sul que venceu," in *Ponta de Lança*, 5th ed., Obras Completas (São Paulo: Editora Globo, 2004), 104–12. Originally published in *O Estado de São Paulo*, September 2, 1949.

69. Luiz Felipe de Alencastro, *O Trato dos viventes: Formação do Brasil no Atlântico Sul, séculos XVI e XVII*, (São Paulo: Companhia das Letras, 2000). A truly native workforce in Brazil would only be constituted in the mid-20th century, after the great immigration waves ranging from the Great Depression of 1873 to the one in 1930, right after the crash of the New York stock exchange. Before that, Brazilian history oscillates between the accelerated speeds

of the South Atlantic and the slow and painful "pacification" of the backlands, a burning issue not only in politics and economics, but also in literature.

70. In geopolitical terms, it is England from the industrial revolution era, and not properly Portugal, that occupies the privileged role of the commercial triangle in the 18th and 19th centuries, grabbing the resulting lion's share of those trades.

71. Negri and Cocco, *Glob(AL)*, 77.

72. Bandeiras (meaning "flags" in Portuguese) were expeditions, generally from coastal or near-coastal strongholds to the backlands, in order to hunt and slave indigenous peoples, prospect natural resources, and expand the dominions of big colonial landlords. They happened regularly from the 16th to the 18th century in Brazil. One of the most common starting points for the bandeiras was the town of São Paolo, on the Southeastern plateau. The white elite members of those expeditions were called *Bandeirantes*.

73. Abdul-Karim Mustapha, "O Paradigma das duas fronteiras do Brasil," trans. Patrícia Farias, *Revista Lugar Comum*, no. 11 (August 2000): 15–22, accessed March 2016, http://uninomade.net/wp-content/files_mf/111712120505O paradigma das duas fronteiras no Brasil-Abdul-Karim Mustapha.pdf.

74. Mustapha, "Paradigma das duas fronteiras . . .," 19.

75. See Karl Marx, "Chapter XXXIII—The Modern Theory of Colonisation," in *Capital: A Critique of Political Economy*, vol. 1 (North Chelmsford, MA: Courier Corporation, 2011). As he puts it: "It is the great merit of E.G. Wakefield to have discovered, not anything new about the Colonies, but to have discovered in the Colonies the truth as to the conditions of capitalist production in the mother-country. [. . .] However, we are not concerned here with the conditions of the colonies. The only thing that interests us is the secret discovered in the new world by the Political Economy of the old world, and proclaimed on the housetops: that the capitalist mode of production and accumulation, and therefore capitalist private property, have for their fundamental condition the annihilation of self-earned private property; in other words, the expropriation of the labourer."

76. Quoting Alencastro, Mustapha presents another indication of the persistence of the biopower bloc and its extreme violence as unquestioned background of Southern development theories: "A Luso-Brazilian, and later, an authentically national method of terror, the shock of the barbarian will of the master—seeking to demonstrate to the newcomer his new subhuman state—was again practiced during the dictatorship of 1964–85. Taught by the long slavery experience, the torturers of DOI-CODI and of Operação Bandeirantes also appealed to beatings at the entrance of police stations and barracks in order to dehumanize and terrorize the suspects of 'subversion.'" Alencastro, *Trato dos viventes*, 148 *apud*. Mustapha, "Paradigma das duas fronteiras . . .," 20. We might add the intimidation, torture, and summary execution practices that are the bread and butter of police operations in urban and rural areas of Latin America, a continent where 42 of its cities are among the 50 most violent in the world, according to the murder rate per 100,000 inhabitants. *Cf.* "List of Cities by Murder Rate," Wikipedia, accessed February 2016, https://en.wikipedia.org/wiki/List_of_cities_by_murder_rate.

77. Fabricio Toledo, "Fuga como resistência: A pobreza criando excedentes," *Revista Lugar Comum*, no. 31–32 (February 2012): 103–13, accessed March 2016, http://uninomade.net/wp-content/files_mf/110410120852Fuga como resistência-a pobreza criando excedentes-Fabrício Toledo de Souza.pdf. See also Sandro Mezzadra, *Diritto di fuga*: *Migrazioni, cittadinanza e globalizzazione*. (Verona: Ombre corte, 2006).

78. Fernando H. Cardoso, introduction to *O Mito da Marginalidade*, by Janice Perlman (Rio de Janeiro: Paz e Terra, 1977), 13–15.

79. Fernando H. Cardoso, *Capitalismo e Escravidão no Brasil meridional: O negro na sociedade escravocrata do Rio Grande Do Sul* (Rio de Janeiro: Civilização Brasileira, 2003 [1962]), 352. Emphasis added.

80. José de Alencar, *Cartas a favor da escravidão*, comp. Tâmis Parron (São Paulo: Hedra, 2008), 86.

81. Negri and Cocco, *Glob(AL)*, 78.

82. Ibid.

83. Alencastro, Cit., 198–99.

84. The first army for capturing indigenous people and conquering territories was formed in São Paulo and consisted of a *caboclo* contingent–cannon fodder for much of the *bandeiras* of the 18th century—renowned for their ruthlessness and brutality, resulting from the vertical miscegenation between the white colonizer and the indigenous people who inhabited the plateau region. *Cf.* Darcy Ribeiro, *O Povo Brasileiro: A formação e o sentido do Brasil* (São Paulo: Companhia Das Letras, 1995), 95–98.

85. Sandro Mezzadra, Julian Reid, and Ranabir Samaddar, *The Biopolitics of Development: Reading Michel Foucault in the Postcolonial Present* (New Delhi: Springer, 2013).

86. *Nota bene*. Biopower and biopolitics are exactly the same. They share an identity of nature as far as there is no point of view external to the web of power relations that could judge "good power" and "bad power." Only from a genealogical or strategical perspective—that is, through practices that can transform the exercise of power—is it possible to distinguish one from the other. The challenge here—as in philosophy with the concepts of "will of power," from Nietzsche, or "*virtù*," from Maquiavel—is to make an immanent differentiation of power. Thus, the difference between the biopower bloc and the biopolitical resistances—for instance, within the miscegenation or migrations—can only be understood as a difference of functioning modes, which have to be referred to the strategic plan and concrete situations to make sense. One must be careful, however, that the opposition does not become itself a criterion that transcends the situations, as if a jurist-philosopher with its adjudicatory will could rule: "This is biopower, that is biopolitics." The whole point to the biopolitical displacement consists in adopting a nuanced approach to the resistances in order to see (and activate) the rupture lines *from* immanence. All this is to point out what the reader has probably already noticed—that is, a certain degree of conceptual indecision between biopower and biopolitcs, which is deliberate and consequential.

87. Ibid., 233.

88. On migration as biopolitical resistance, see Leonora Corsini, "As migrações e o trabalho da resistência," *Revista Lugar Comum*, no. 19–20 (July 2009): 211–21, accessed March 2016, http://uninomade.net/wp-content/files_mf/113003120838As migrações e o trabalho da resistência - Leonora Corsini .pdf.

89. Alencastro, *O trato dos viventes*.

90. According to Abdul-Karim Mustapha, the historical and political mosaic of Alencastro may be productively supplemented, through a *positive approach*, with the proletarian formation theory from the slavery perspective, by Yann Moulier Boutang. *Cf.* Mustapha, "Paradigma das duas fronteiras . . .," 19–22.

91. On the relation between smooth and striated space, see Gilles Deleuze and Félix Guattari. "Plateau 12—1227: Treatise on Nomadology," in *A Thousand Plateaus: Capitalism and Schizophrenia*, trans. Brian Massumi (London/Minneapolis: Minnesota Press, 1987 [1980]).

92. Curiously, Slavoj Žižek classifies Canudos as an "exemplary case of such 'a return' of the repressed" in Christian tradition. *The Courage of hopelessness*, note 20, p. 297.

93. The backlands eventually turned into sea thanks to the developmentalist project of the dictatorship (1964–1985), with the Sobradinho Reservoir at the São Francisco River basin, to which Conselheiro referred. In the 21st century, however, the environmental destruction caused by the developmental intensification completed the cycle, when the sea turned back into backlands, as the river dried up.

94. Euclides da Cunha, "Chapter II: Man," in *Backlands: The Canudos Campaign*, trans. Elizabeth Lowe (New York: Penguin Classics, 2010 [1902]). See "A Backland Hegira," in "Subsection IV—Antônio Conselheiro: A Misfit Turns."

95. Curiously, the millenarianism of Antonio Conselheiro has self-proclaimed monarchist, projecting an image as Lusitanian as São Sebastião in the possible return of the Emperor Pedro II, who was in exile in Portugal. For the blessed of Canudos, the Republic meant nothing but the Kingdom of Antichrist. On rebellions and uprisings in the transition from Empire to Republic (1889), see Martine Droulers, *Brésil, une géohistoire* (Paris: Presses Universitaires de France, 2001), 115.

96. Paul Gilroy. *The Black Atlantic*. Modernity and Double Conscioussness (Cambridge: MIT Press, 1993). *O Atlântico negro: Modernidade e dupla consciência*, trans. Cid Knipel Moreira (São Paulo: Ed. 34, 2001).

97. Peter Linebaugh and Marcus Rediker, *The Many-headed Hydra: Sailors, Slaves, Commoners, and the Hidden History of the Revolutionary Atlantic* (Boston: Beacon Press, 2000).
98. P. Gilroy, cit. 4–5.
99. We are mostly thinking of *Black God, White Devil* (1964), *Entranced Earth* (1967), and the splendid *The Age of the Earth* (1980).

Chapter Two

The Birth of Real Neoliberalism

The 1980s and 1990s became known as the long neoliberal night. Since the following decade, any theoretical or political discussion regarding the Latin American Left-leaning regimes invariably considers that period the primary negative reference for changes. Being alert for the comeback of neoliberalism became a watchword for the Left, the progressives, and the nationalists in general, who constantly warn against the risk of historical regression, presenting the image of a nightly ghost that haunts the achievements and advances of the present. We must be careful, though, so that the night is not so dark as to turn all cats gray. This would mean losing the ability to consider the nuances of political-economic processes that mark that properly neoliberal period in the South, and losing all cutting lines, continuities and discontinuities, in relation to the following "progressive" governments.

Neoliberalism in Latin America was not a reflex of the structural transformation of Capitalism coming from the North. Strictly speaking, its first institutional conscious experience actually took place in post-coup Chile, along with the murderous repression of social movements and Left militancy, as well as the dismantling of popular organizations, such as autonomous health centers in poor areas. After 1973, the dictatorship of General Pinochet imported economists from the Chicago School, whose star was Milton Friedman. The Chicago boys, as they have come to be known, received *carte blanche* to outline the reorganization of the local productive system. They applied an economic policy of capitalist shock with an abrupt free-trade opening, the dismantling of existing institutions of social security, and repeated administrative, budget and financial adjustments in order to launch Chile as precursor of the new global economy reality.[1] The ECLAC structuralism and the import substitution model of the 1950s were definitely buried, considered to be no more than archaisms of the second post-war period. It no longer worked under the new conditions. In the 1970s, export-led growth would take over Chile, replacing the previous model, with the exception of the vital copper mining sector. In this sense, during the 1970s, Chile was a laboratory for the governmental turn that later would spread around the world, especially with the Reagan era. It is ironic—in fact, it is suggestive—that an originally nationalist and technocratic government has been the first to introduce the neoliberal *and global* revolution of capitalism as economic policy. Chilean's neoliberalism was far from operating by the "invisible hand" of the market, depending otherwise on the iron hand of the State.

Another early experience of neoliberalism in the region took place in Argentina during Videla's and Gualtieri's dictatorship (1976–1982), with the liberalizing reforms carried out by Minister Martinez de Hoz. In a hyperinflationary context with annual rates up to 5,000%, they adopted a shock therapy similar to the one in Chile, implementing tax and administrative reforms aimed at reducing the state budget and privatizing state-owned sectors, opening trade, embracing monetarism, raising interest rates, and increasing the dependency on foreign capital. Meanwhile, in the background, labor struggles and social resistances were crushed through wage restraints, cruel and bloody political persecution, and direct organizational interdiction. According to their own terms, more successful in Chile than in Argentina, the early reforms ended their first cycle in 1982, with the economic collapse of Latin American countries due to debt and the defeat of Argentina in the Falklands War. The inability to gain return in productivity and competitiveness compatible with the debt obligations, following a world crisis of capital adjustments, led to a chronic insolvency situation, forcing the national States to appeal again to international capitalist institutions. The result was a decade of economic stagnation, real wage acute decline, and hyperinflation outbreaks. During the so-called "lost decade" to Latin American economies, a second wave of institutional neoliberal reforms much more pragmatic than the "purer" reforms inspired by Friedman in the mid-1970s, began. Organizations like the International Monetary Fund (IMF) and the World Bank (WB) assumed the leading role, and their credit injections came associated with the imposition of budget restructuring programs. Historical highlights of this new phase of neoliberal Latin America are Minister Domingo Cavallo's plan, during the Argentinian government of Carlos Menem (1989–1999), which privatized companies and introduced the currency board;[2] the Washington Consensus–oriented reforms of the second Venezuelan government of Carlos Andrés Pérez (1989–1993), with the unfreezing of gasoline prices, which is one of the destabilizing factors that led to the great Caracazo uprising (1989); Fujimorism in Peru (1990–2000); and the structural adjustment measures carried out in 1985 by the octogenarian Bolivian President Víctor Paz Estenssoro.

In the Western world, the neoliberal revolution hammered the final nail in the Keynesian one, which came before it, and in a "development economy" that, after all, has found some political-theoretical legitimacy in the first one.[3] The struggles of the long curve of 1968, understood in a broad sense, charged global capitalism with such unbearable pressures and demands that it was compelled to take a new restructuring leap. Once more in its history, capitalism had to adopt a new configuration to face the challenges posed by labor mobilizations.[4] It is this turning point that Foucault traces through a genealogy of power in *The Birth of Biopolitics*—this, by the way, was a subject also discussed by a plethora of thinkers of the "great transformation" occurring between the 1970s and 1980s.[5] One of the milestones for this restructuring was the Nixon shock of 1971, which represented the final blow against the international financial system of Bretton Woods, a milestone toward the flexible, fluctuating, venturesome era of financialized governance of globalization. That same year marked the beginning of privileged relations between China and the United States as well as the outbreak of a military effort called the "war on drugs," whose aim was Latin America. The US president suspended the dollar to gold convertibility once and for all, eliminating the need for backing when issuing currency. Other milestones were the two oil shocks, in 1973 and then in 1979, caused by the cartel producers' reaction to low prices and the geopolitical reconfigurations in the end of the Cold War period, making energy costs more expensive overnight, having a direct impact on every industry in the world. The oil shortages aggravated the global economic crisis, leading to "shock" solutions also on the institutional level. After all, the main target was now to dismantle the welfare state guaranteed by the Fordist-Keynesian framework in order to constrain labor into a new configuration of exploitation, more suitable to the post-industrial accumulation regime and financialized governance.

Nevertheless, limiting the criticism of neoliberalism to the dissolution of welfare, financial deregulation, production out-sourcing, and job instability would remain on the surface of the phenomena. Beyond capitalist restructuring lies the fringe of subjectivity transformations, which it must contain and exploit in a reflex action. Hence the importance of extrapolating analysis in terms of institutional renewal, so we understand the most significant changes of social practices and production relations. The turn that led to neoliberalism consisted in a transformation, at the capitalist core, of the very driving force that determines value and wealth. That alters not only the regulation of the economic and financial system, but also the nature of labor, social cooperation and the exploitation methods.[6] Social productivity becomes innervated by a matrix of networks, and capitalist value depends on affective, cognitive, and care qualities of immaterial labor. Thus, the global capitalist restructuring demands the political and economic rebuilding of a regulation capable of gathering minimum efficiency. Therein lies the micro and macroeconomic formulation of post-Fordist regulation—which Foucault calls neoliberal governmentality and Deleuze, societies of control—that assimilates social dynamics as a gas, saturating a world of productive flows and networks that threaten to flee from all sides.[7] Throughout the 20th century, the factory was the paradigmatic production unit of capitalism, which corresponded to the Fordist regulation. The factory unit was then reproduced throughout society as a whole, making every cell a small society-factory: the family, the school, the office, the hospital, the jail, public institutions, and so on. According to the neoliberal theory, the industrial society was modeled as a large factory, with its departments, arborescent organizational charts, and linear productive chains—a dream Henry Ford shared with real socialism since the 1920s. After the global economy restructuring of the 1970s, in the new post-Fordist regulation, the reference becomes the "social factory,"[8] a biopolitical fabric entirely invested with productive technologies. So, instead of a "productive sector"—usually associated with industrialization and, within it, capital goods—there is now a diffuse social productivity that, in the global context, exceeds the surplus accomplished in Fordism. The productive sector is intertwined with the very lives of populations, in their immanent productive force, generating a surplus value of life. It becomes imperative to control it.

In *The Birth of Biopolitics*, Foucault describes the theoretical effort of neoliberal economists to develop a program, an institutional framework that could achieve this task "from inside and from below." Their immediate professed enemies were Nazism, Stalinism, and the Western welfare state, but, regardless of those ghosts, they were actually anticipating the urgent need for a flexible governmentality that capitalism would have some decades later. These theories formulated by German ordoliberals and anarcho-liberals from Chicago, as we have seen, are not guided by the mere dissolution of the state and deactivation of power, according to the logic of "minimal state," giving way to the market logic. That would mean a return to the classic tradition of political liberalism, which neoliberalism is definitely not, even though critically, claiming a "free market," totally independent from the state hand, would mean accepting the premise that market and economy could operate under an autonomous dynamic and stay clear of state regulation. The neoliberals are, in fact, aware that there is no market without state, that market and state logics focus on the same problem of how to handle the crisis. That is why those theories cannot be reduced to a position regarding a mystified game between state and market, as if those two heavy words have now the same meaning they had in Smith's or Keynes's time. In fact, as Foucault describes it, neoliberal formulations aim at guiding a change in the nature of power exercise. This change corresponds precisely to an answer to the drives of capitalist restructuring. That explains the monetarist, diffuse, "gaseous" regulation based on the corporatization of society and on the *homo oeconomicus* as the figure of subjectivity which will saturate post-Fordism beyond the moral of the Fordist workers, the good old employers and employees. The new social factory is no longer organized into departments and staggered management cycles, but based on entrepreneurship and the corporatization of individuals and their relations in multiple aspects. The prototype for the social relation of capital is not the factory job with fixed income anymore, but the link between service and a molecularized salary, which comes from many sources: projects, jobs, qualifications. It is a great tectonic movement, in part anticipated by the neoliberals from Chicago. The "value of values" is not full employment anymore, as in Keynes's macroeconomic recipe, but full activity, while GDP growth is no longer necessarily indexed to employment rates.

Neoliberalism produces new molecular institutions, much more disseminated throughout the social fabric, which recognize the biopolitical dimensions of social production. They recognize these dimensions because they have no alternative, because they need to control them and learn how to deal with them from the perspective of exploitation.[9] In this sense, finance is no longer a separate capital, in an external relation of alliance with industrial capital as before, and now invests in life itself and financializes social relations. The financialization of life "from below" and "from inside" participates in the restructuring that led to neoliberalism as one of the main mechanisms of flexible control.[10] At the same time, the economization of society is driven by labor in its entrepreneurial mode.[11] With the post-Fordist turn, the states become actors amid a scenario now occupied by multinational institutions and corporations, financial system organizations, and delocalized global networks that connect people. We are at the age of new capitalist globalization, which Antonio Negri and Michael Hardt called *Empire*.

The old fiscal and administrative mechanisms, the old statist-Keynesian or national-developmentalist precepts, they simply do not work anymore, they cannot cope with the transformations. And this inability is because the living work and the nature of capitalist value have undergone an irreversible change, which in turn is immediately ambivalent. The capital process was restructured to capture these transformations, in accordance with the biopolitical governance; at the same time, the social composition changed to give life to struggles of a new kind. The neoliberal governmentality is permeated, from the beginning, by this antagonism, this internal pressure that no longer works according to a march in which every struggle converges in a single fight. Instead, it operates multiple struggles that exert counter-power from the very molecular fabric in which they are intertwined.[12]

At this point, we could ask: Why in Latin America first? Why does Chakrabarty's formula "first in Europe, then elsewhere" seem inverted when it comes to neoliberalism? Real neoliberalism was born in Chile and Argentina during the 1970s, restoring class power after military coups were backed by local and global elites. It then spread throughout Latin America in the

1980s, at the same time it became a worldwide tendency under Thatcher's and Reagan's administrations. Was it just a dress rehearsal for the real thing that was being applied in the North? We don't think so. Again, we could mention Marx's remarks on E. G. Wakefield, in the 33rd chapter of *Capital*, about the colonized periphery revealing the "secret" of the capitalist core functioning and the so-called primitive accumulation. As David Harvey puts it, "[n]ot for the first time, a brutal experiment carried out in the periphery became a model for the formulation of policies in the centre."[13] As a matter of fact, when it comes to real neoliberalism, there is an internal reason why Europe was "provincialized" by Latin America and not the other way around.

The fact that the "shock moment" of coups, torture, and expropriation at the *pudenda origo* was successful in restructuring capitalist governance of globalization does not imply that a "colder," more distended temporality was not building up at the same time. In order to work in the long run, primitive accumulation must be reproduced on a daily basis, and its flame must not die out at the origin so that appropriative violence can be inscribed in social relations as a means of continuous exploitation.[14] That is how capitalism produces its own linear history (*h1*), a temporality of dominance that depends directly on mechanisms of measure, discipline, truth regimes and routine practices. Marx meets Foucault and primitive accumulation becomes governmentality, a power organization that guarantees regularity and efficiency to the workings of capital.

As Pierre Dardot and Christian Laval's book on neoliberalism indicates, neoliberal successful introduction depended also on "protracted, continuous, often silent war, which is irreducible to the 'shocks' that serve as a pretext for some particular offensive."[15] Governmentality cannot be reduced to a violent coup of the elites that would come from top to bottom. Instead, it is a steady reaction of capitalist powers, from top down *and* bottom up, that ends up affecting the society as a whole, the entire operation of productive social relations. As Foucault explained in his lessons of 1978–1979, this is the only way the new regime of power and truth could exist in the first place, since the neoliberal distinctive aspect is to operate *from within*, introducing itself into civil society, in its very constitution as "civil." Neoliberalism is not just a macropolitical program marked by cuts in public spending, privatizations, monetarism, and financial globalization. It cannot be limited to a shock doctrine prescribed by think tanks and applied by financial institutions backed by capitalist powers.

This reductionism appears clearer after the electoral successes of Brexit (in the United Kingdom) and of Trump (in the United States) and it is confirmed also by the kind of defeat of Marine Le Pen by Emmanuel Macron in France. On the one hand, it is difficult to see these electoral victories as continuities of neoliberal politics; on the other hand, Macron defeated Marine Le Pen by presenting himself as an outsider of the traditional party system. We can agree with Naomi Klein when she writes that "even if this nightmarish presidency (of Donald Trump) were to end tomorrow, the political conditions that produced it, and are producing replicas around the world, will remain to be confronted,"[16] but we cannot simplify or reduce the emergent conflicts between "globalism" and "new sovereign" approaches. In the same way, it is really useful to look at the catastrophic neodevelopmentist experiences during the pink tide in South American governments. The ideological critics of neoliberalism, even when they are red or pink colored, brings us to even more closed impasses.

We should remember Deleuze and Guattari's warning in the 9th plateau, called "Micropolitics and segmentarity," of their 1980s *magnum opus*:[17] "[E]verything is political, but every politics is simultaneously a *macropolitics* and *micropolitics*. Take aggregates of the perception or feeling type: their molar organization, their rigid segmentarity, does not preclude the existence of an entire world of unconscious micropercepts, unconscious affects, fine segmentations that grasp or experience different things, are distributed and operate differently."[18] Neoliberalism works both on a global level, through major institutional processes, which extend to all economic spheres, as well as by the most pervasive molecular means that spread everywhere, like propagation waves. This is precisely the dual path that financialization brings together when operating within life, meaning it couples supranational networks of banks, institutions, risk agencies, and organizations with internal mechanisms to social relations of credit, investment, debt, and consumption. As Dardot and Laval conclude: "It is therefore essential that we understand how ordinary, routine violence is exercised today, rather like Marx when he observes that the domination of capital over labor only exceptionally involves extra-economic violence; that it is more commonly exercised in the form of a 'silent compulsion' inscribed in words and things."[19] The machinery of neoliberalism is not limited to unregulated markets (economicism) or ideological hegemony sustained by superstructural domination (politicism). There is a second temporality composed of subjective productions, which Marx called the "secret" of primitive accumulation, that conditions those political and economic manifestations of neoliberalism in first place. Neoliberalism finds a propitious social environment in this second temporality, impregnated with micrological spaces of behaviors and assemblages, and intermingles with it, breaking loose the wires of a particular power scheme. That is the South anomaly within capitalist history that real neoliberalism reveals.

Neoliberalism works within a logic conceptualized by Deleuze and Guattari as "polymorphism."[20] As they describe it, capitalism adopts an isomorphic approach when it deals with systemic centers, but employs a more complex operating system, involving multiple irreducible problems, at peripheral borders, between centers and the third world, thus allowing primitive accumulation to normalize continuous exploitation. In other words, this enables capitalism to succeed in subsuming society. In this approach to world economy, the "third world" is not an Outside that could organize itself autonomously and relate to capitalism as an alien rationale, as some uneven development and dependency theories present it. For Deleuze and Guattari, "[w]hen international organization becomes the capitalist axiomatic, it continues to imply a heterogeneity of social formations, it gives rise to and organizes its 'Third World.'"[21] What is more, this does not occur only between Western nation-states and the rest of the world, since the relation center/periphery is stratified: "in these formulas the South is an abstract term designating the Third World or the periphery, and even that there are Souths or Third Worlds inside the center."[22]

So, the authors of *A Thousand Plateaus* are not reproducing another dependency or imperialism theory. On the contrary, they say dependency relations "are not the most important" issue in today's capitalism. Their conceptualization of world market would be more in tune with interdependency theories, such as Cardoso's. According to Deleuze and Guattari, "[t]he international capitalist axiomatic . . . tolerates, in fact it requires, a certain peripheral polymorphy, to the extent that it is not saturated, to the extent that it actively repels its own limits; this explains the existence, at the periphery, of heteromorphic social formations, *which certainly do not constitute vestiges or transitional forms* since they realize an ultramodern capitalist production."[23] The subsumption to capitalism of the non-abstract South poses different problems than the ones raised in central economies and introduces different ontological perspectives for governmentality to deal with. Due to the coexistence of properly capitalist and heteroclite relations of production, capitalism has to go through a metamorphosis so its logic—"axiomatic," as the authors call it—can work effectively. It is as if the capital process has the internal capacity of restoring polymorphism to its own structural purposes.

In other words, neoliberal capitalism is the most adequate regime of accumulation to subsume a polymorphic society in both cases: when the polymorphism is the product of social diffusion of the Fordist and disciplinary fabric, or when it is the heritage of colonization and under-development. Here we can agree with Žižek, when he says that "global capitalism has no problem in accommodating itself to a plurality of local religions, cultures, traditions. So the cruel irony of anti-eurocentrism is that, . . . global capitalism no longer needs Western cultural values."[24]

Extending the argument, we could say that the Southern polymorphism of global capitalism requires that its axiomatic rationale operates according to a different criterion. Instead of falling within the logic of *transition*—from precapitalism to capitalism, from archaic to modern relation modes, and so forth—it works based on *translation*, which presupposes the coexistence of multiple temporalities. Abandoning the *transition* logic in favor of the *trans-*

lation process also means cutting by the root every analogy that socialist formulations try to convey to developmentalism, which disregard the polymorphism involved. The change is imperative for capitalism to really succeed in subsuming the heterogeneous and productive temporalities of *h2s*, if we are to use Chakrabarty's grammar.

This displacement from *transition* to *translation* has nothing to do with accepting a new kind of multiculturalism; on the contrary, multiculturalism is an *effect* of *h1*. However, it is important to remember, that the *h2* lines are not deactivated or "killed" along the process, remaining alive as tension-limits. That is exactly why the logical outcome of capitalism in peripheral borders is called polymorphism instead of isomorphism. Thus we could say that neoliberalism is another name for the translation process necessary for subsumption (i.e., normalization of primitive accumulation). From the subjectivity production perspective, neoliberalism means a reconfiguration *en abime* of social relations that cannot be "isomorphized" without losing their singularities (as *h2s*). The neoliberal governmentality has to deal with a multiplicity of productive and living material logics that do not share any unifying transcendent value, a frame of reference for commensurability or a reasonable common denominator, a reality that cannot be totalized: neoliberalism is the art to govern by the continuous translations of social flux that no one can anymore organize along discipline lines. From the point of view of governmentality, it is the diagram of power that has to deal with polymorphism, subjective heterogeneity, and full incommensurability, and this is the grounds for neoliberal financialization.

Scottish sociologist Andrew Ross provides a good example of how neoliberalism is able to translate heterogeneous subjectivity processes into capitalist enterprise projects, sometimes by the means of labyrinthine trajectories and *quid pro quo*.[25] He develops the case of contemporary China, where the party-state has decided to strategically catch up with regard to *creative industries,* aiming at the high aggregated value of the post-industrial circuits. At the same time, the party-state, faithful to Deng Xiaoping's maxim that Chinese economy should "cross the river by groping the stones," cannot afford giving too much cultural freedom to the potential creative networks. Surely, the Communist government did not forget the political effervescence that led to the multitudinous Tiananmen protests in 1989. As Ross recalls, the party's leaders have classified these protests as "cultural excess," a late outburst of the old Cultural Revolution. With that in mind, Chinese policies regarding the creative and cultural turn are being enforced with caution in order to avoid *detours* from the controlled trajectory desired by the central authority. This explains why the Chinese government has been promoting traditional, even folkloric aspects of national culture and extolling Neo-Confucianism, going the exact opposite direction of Mao's Cultural Revolution of the 1960s, which was violently against the past and anti-imperial traditions par excellence. Interestingly enough, Ross traces a genealogy of neoliberalism from a cultural point of view, which goes from the "extremely idealized" reception of the Cultural Revolution in the West until the creative industries program launched in the 1990s by neoliberal Anglo-Saxon governments. Once it percolated through the leftist youth culture during the 1960s and 1970s, Maoism determined the attitude represented by slogans like "to rebel is right" and "bombard the headquarters," which promoted self-expression and defied every institutional framework and existing discourse. "Talking bitterly" at places of authority became a trademark of an entire generation of collectives and new social movements. The Maoist reception conditioned a behavioral style connected to second wave feminism, black identity movements (such as the Black Panthers), and all sorts of fierce attacks against the straight white male cultural dominance, both in the institutional and personal relations level. Then, in the 1980s, this helped spark the Anglo-Saxon "culture wars," between a cry for reforms and for the reeducation of every institutional framework (school, office, gender, family, etc.) and the conservative or liberal counter-criticism impregnated with "political correctness," moralist self-righteousness, and sectarian tendencies. From the Cultural Revolution to the "culture wars," Ross argues that, in the 1990s, we finally reached a new technology of power related to a new management style and more horizontally organized enterprises. In one scene from the movie *Capital* (2015), by Greek director Costa-Gavras, we see an example of the assimilation of Maoism by the "new spirit of capitalism" when the bank president decides to revolutionize the institution (and thus seize power for himself) by calling all employees to "bombard the headquarters," denouncing managers and supervisors. According to Ross, "Maoist cultural policy had more to do with the transformation of subjectivity and the reeducation of citizens,"[26] precisely what neoliberalism pursues in order to produce the self-entrepreneur. In the end, the Chinese attempt to emulate creative industries during the 2000s is actually the culmination of a long line of detours, from the Cultural Revolution to the Neoliberal Cultural Turn in the Asian country. This kind of investigation conducted by Ross, which examines the development of figures of subjectivity and their heterogeneous transformations,[27] is much sharper in tracing a neoliberal subjective cartography than those who simply diagnose a "neoliberalism with Chinese characteristics" through macroeconomic policies or big institutional framings.[28]

In *La Razón Neoliberal*,[29] Argentinean researcher Veronica Gago repeats, 30 years later, the analysis of Hernando de Soto in his *El Otro Sendero*,[30] but in a situation dominated by the neodevelopmentist tentative during Kirchner's government. Therefore, she describes actually existing neoliberalism operating in popular markets and informal networks, their figures of subjectivity and circuits that extend to income, credit, and consumption flows. Having researched the *La Salada* market at the outskirts of Buenos Aires,

Gago distinguishes a "neoliberalism from below," marked by a complex set of practices to which no easy formulas or possible dichotomies correspond. Legal or illegal, exploitation or resistance, blackmail or consent, there is no way to describe that field's real functioning by appealing to polarities. She faces a frontier zone where everything seems to be thresholds, gradients, and intensities, which could only be properly assessed in terms of immanent strategies. It is precisely there that molecular neoliberalism operates with its governmentality based on self-entrepreneurship, the economy of illegalities, pervasive subjective debt, intensified competitiveness, and systemic drive for efficiency, to quote some of the Foucauldian terms mobilized through the analysis. For Gago, beyond any simplistic cleavage between the 1990's real neoliberalism and the so-called post-neoliberal progressive governments of the 2000s—usually thought of in terms of "return to the state" or "growth with social inclusion"—it is necessary to analyze how those molecular mechanisms, related to popular credit networks, consumption growth, differential inclusion, and better access to educational assets and knowledge affect one another and the whole system in its operative unity. The transition from the 1990s *stricto sensu* neoliberal governments to the Kirchnerist years (2003–2015) would be better explained as a metamorphosis of the neoliberal governmentality than the actual genesis of a "post-neoliberal" period. In this sense, as sharply pointed out by the author, neoliberalism is far beyond the think tanks' formulations in tune with the Washington Consensus and far behind "what a good share of the local progressives are willing to understand."[31] The progressive governments' policies do not come from Outside to alter the otherwise existing equilibrium of neoliberal civil society. Instead, those policies (social assistance, popular credit, income policies, and so forth) affect the disequilibrium constituent of neoliberal governmentality from *inside*, which obviously produces an impact on the involved subjectivities at the strategic level. The entire book of Gago is an ardent effort to name the fleeting surplus that eludes neoliberal mechanisms of subjectivity rationale and control, "a certain monstrosity"[32] that exceeds Foucault's diagnosis concerning the neoliberal *homo economicus*. And she attempts to do so without forging another typically Latin American character based on the "good people" notion, refusing both the common sense—like the idea of a Brazilian *povão*—as well as more theoretically sophisticated populisms.[33] Building upon the Ecuadorian philosopher Bolívar Echeverría, Gago strives to give consistency to the vitalist popular forces organized into strategies and networks which constitute tension-limits, resulting in potentially antagonist focal points *within* neoliberal governmentality. She creates the notion of "Baroque economies" and "popular pragmatisms" in order to bring the analysis closer to the real problems of confronting neoliberalism "from below." It is a sort of *anthropophagical descent*, to speak in the Modernist Brazilian writer Oswald de Andrade's terms,[34] a necessary movement to put the feet on the ground where power is exercised on a day-to-day basis and not on some cloud made of idealistic notions.

Francisco de Oliveira, as soon as 1972,[35] had already identified that the massive informal economy in Brazil, especially in big cities, was not a remnant of non-development, a non-capitalist space "in the waiting room" of the history of capitalism. Oliveira was challenging national developmentalist economists, such as Celso Furtado, who sustained that mass industrialization with a strong internal market would be the solution for the underemployment problem in peripheral capitalist societies. In fact, as he demonstrates in elegant lines, the industrialization of the South, for its very peripheral condition, was developing the informal sectors, not bumping into them on the way. Under political and economic coordinates, determined by internal and external factors, the new approach to development was forging dualisms all over the process, as it happened in Brazil during the long developmentalist cycle from 1930 to 1980, since the Vargas Era until the end of the military dictatorship. That is the history of a promethean nightmare that reproduces the same barriers and vices which interventionism was supposed to solve in the first place. After all, informal sectors were never pre-capitalist spaces that development would eventually overcome, since the very national development aimed at industrialization created a new informal sector, inflating the participation share of services in the economy. However, Oliveira does not echo the pessimist "development of underdevelopment" thesis (of Gunder Frank and others). Although he agreed that catching up with a full employment society in Fordist-Keynesian lines, as dreamed by the national developmentalist, was absolutely chimerical, he also understood that it would be possible to modernize specific economic areas at the expense of others and, more importantly, he believed the power of class organization could push into other political directions. In any case, Oliveira's analysis leaves a blank spot concerning the production of subjectivity *within* the informality basin, popular or service economy, and precarious labor composition so typical in the South, which he condemns and disregards as a disorganized broth without class strength. Albeit lacking the subjectivity point of view, Oliveira's analysis of the "strange world of dualism" is very topical. As he successfully employs the "integral method"[36] by covering economics, the political moment, internal and external market factors, and sociological aspects, his theoretical work is able to offer a complete panorama of the capitalist accumulation regime, whose final objective is the production itself, and its historical transformations through present Brazilian development policies. He demonstrates how nationalist industrialization and technical progress reproduced and reinstated the expansion conditions of the productive system. However, he does not address the effects of transitioning from an agrarian and export-led society to an urban and industrialized society through the viewpoint of new social and political *agencement* (we decided to keep the French word because of the ambiguity of English translation: assemblage is the result—ex post—of fragments and parcels combination, but *agencement* is the *process* of combination itself, beyond totality and fragmentation. The informal and service sectors, the fragmented metropolis, the fluid dynamics of social productivity, all that seems mere post-ideological liquid modernity with a strong subjective dimension. From Oliveira's point of view, no class power could spring from this disorganized soil cultivated by capitalist process in the South.

We think that this very lack of structure—persistent both in uneven development and dependency theories as well as in Marxist critiques like Oliveira's—is a qualitative aspect of the South. What we can do to exit from his impasse is to modernize the modernist concept of anthropophagy.[37] It is what we tried to do discussing the approaches of Brazilianization as a positive element of subjectivity,[38] opposing it to the new double concept: the becoming-Brazil of the world and the becoming-world of Brazil. In contrast with the negative interpretations of globalization as essentially the world's precarization of labor and *favelização* (turning all into slums, as one can read in a mainstream book about geopolitics: "25 percent of Brazilians are thought to live in the *infamous* favela slums"[39]). The becoming-Brazil of the world and its pair, the becoming-world of Brazil, provoke a symmetrization of perspectives[40] in order to turn post-colonialism and post-modernism into political machines. It actualizes the field of *biopolitical resistances*—which embodies a genealogy full of synchronicities and intensive dimensions in the *longue durée*—to the age of neoliberalism. This then opens the possibility for strategic thinking and actions that go beyond the rigid modes of existence proper to the old Fordist-Keynesian world—that is, the People, the Proletariat, the Party. Building on Eduardo Viveiros de Castro's ontological perspectivism,[41] the symmetrization mechanism does not eliminate differences. In contrast to every multicultural approach, this operation presupposes a subject-*other* instead of assuming the other as subject. The emphasis is on the fact that the *other* is not like us, another subject in the same world, but an Outside from whom we are forced to differentiate. This presents us with other possible worlds, other practical problems interfering with our own. Hence, becoming-Brazil does not mean recognizing two different cultural totalities that could connect through an intersubjective relation, but two sets of variations that affect one another: rather than a relationship between subjects that varies, this implies two poles of variations that relate to each other. Instead of looking for solutions to our problems in the other, the question of becoming involves touching other problems that displace ours and create new ones. We thus wish to extend this "perspective of perspectives" in order to couple the analysis of the transformations in globalized capitalism with the necessary theory of subjectivity, internal to those changes.

THE BRAZILIAN CASE OF REAL NEOLIBERALISM: BETWEEN PRAGMATISM AND INTERDEPENDENCY

In Brazil, the second neoliberal wave arrived in the 1990s with the first directly elected president after the dictatorship, Fernando Collor (1989–1992), and remained throughout the two terms of Fernando Henrique Cardoso (1995–2002) at the head of the federal government. Cardoso's presidency presents a singular opportunity to compare his rich intellectual production—especially after his 1969 important book, *Dependency and Development in Latin America* (coauthored by Enzo Faletto)—with his actions at the helm of the Brazilian government. This allows us to observe how a formerly Marxist development theorist became leader of the reforms that put an end to the last sediments of the dictatorship's national developmentalism and inscribed the Brazilian economy into the flows of global post-Fordist capitalism. Nothing is crystal clear in this trajectory.

First of all, neoliberalism in Brazil has always been marked by a historical ambiguity that does not allow ideological oversimplifications. This relates both to the dual character of post-Fordist restructuration and the overcoming of barriers regarding the economic, political, and institutional structures with heavy developmentalist traits inherited from the military regime (1964–1985). Any critique that reduces the 1990s to the homogeneous predominance of the Washington Consensus is running over a much more syncretic functioning, missing internal and external articulations of the new productive matrix. The turning point of Brazilian neoliberalism occurred at the beginning of Cardoso's first term and even earlier, when he was Minister of Finances of the temporary presidency of Itamar Franco (after the impeachment of Collor) and implemented the *Plano Real*. The plan, which instituted a new currency, the Real, was a set of macroeconomic, financial, and fiscal measures that led to the containment of an endemic inflation permeated by hyperinflation outbreaks and actions for currency stabilization. For over 20 years, the Plano Real stands firm as the macroeconomic framing of all governments, including (or especially) the 13 years the Workers' Party stayed in office (2003–2016). Besides implementing the institutional mechanisms in order to face chronic inflation rises, the Cardoso government's program also provided the needed currency stability which would enable the coming struggles for re-appropriation of income, later crystallized into social distribution policies.

In 1995, when Cardoso took office for the first time, he announced that the populist Vargas era was dead.[42] But he was referring to more than just Varguism. His *bellum internecinum* (total war) covered the entire national-developmentalist institutional heritage built during the 1930–1980 period, from the Estado Novo to the military dictatorship, going through the populist democratic governments of the 1950s, the peak of "ECLAC thinking." Cardoso was taking the old dualist scheme seriously, but now he seemed determined to follow the capitalist lines of restructuring without any further delay. The modernizing impulse would not allow any more space for archaisms and bureaucratic-corporate elements, which were accused of being the major national evil by home-born neoliberals. This explicit political resolve of Cardoso's government should not be confused, however, with an adherence to classical political liberalism according to the old dichotomy that simply repeats anti-market and anti-state diatribes.

In his texts from the 1970s,[43] Cardoso already pointed to the idea that the concept of dependency, rather than being overcome, should be reconfigured according to changes in the scenario, such as global financialization, the constitution of multinationals as main productive vectors, as well as the prominence of international credit agencies and investment banks. Against the dependentist, anti-imperialist or ECLAC substitution import model, but also against central planning, labor discipline, and state-centered industrial policy as in the dictatorship years, he had written that development was not incompatible with the new situation of globalization. The concept of dependency should be actualized in order to cope with the transformations capitalism was undergoing throughout that decade. It was necessary to think in terms of a "new dependency," or rather, a complex *interdependency*. Of course this did not mean that the international plan of interdependency had eliminated hierarchies: asymmetries would be restored to the process, although *in other terms*. After all, that new situation did not imply the emergence of a smooth capitalist space of global proportions, but it would be a dogmatic myopic failing to recognize the need for different strategies that could match the striation of the new governance within globalization that would be nothing more than shortsighted dogma. Cardoso's articles from the 1970s already spoke of new opportunities posed by deterritorialized capital flows that could be seized for the benefit of Southern economies.

However, following the very "integral method" advocated by the new president when he wrote his works on development and dependency, one might ask which internal factors were or could be connected to the new configuration of a global market. Cardoso's government did not ride the wave of new movements that came into play in the late 1970s, such as the landless movement (called Movimento Sem Terra, or MST), the new noncorporate unionism, the minority struggle of the Black, feminism, and other movements guided by autonomy like the neighborhood associations for laying out pleas. Instead, Cardoso's plan was made possible by an alliance among eminently urban modernizing elites, which felt represented by the liberalizing social democracy headed by Cardoso, and the old oligarchic-patriarchal sectors, the most retrograde ones in the country, connected to large estates, the extractive industry and the exploitation of labor in rural areas. This glaring contradiction, which confirmed the ever renewed curse of dualism, rather than crippling the government functioning, endowed it with a correlation of forces capable of applying the changes. The eternal dualism between archaic/agrarian and modern/industrial, which marks the structuralist analyses, once again showed its true face: a broad capitalist alliance against the constituent power of the multitudes. Indeed, pressured by the long arc of struggles increasingly stronger since the 1980s, perceiving a dreaded nemesis in the image of the unionist leader Lula as candidate, the major landowners had no other option than the "escape forward" strategy, falling into the neoliberal project.[44]

The result of this arrangement was twofold: if, on the one hand, the opening to foreign markets and the shrinking of the industrial policy crippled a wide range of industries without competitiveness and productivity, on the other, the government's interest policy ensured compensation to those capitalists who had no way to face the new conditions—their withdrawn capital could very well survive on the profiteering made possible by high interest rates.[45] Without being able to count on the direct strength of the struggles, it is as if the neoliberals were compensating the total losers of the historical process of modernization. Cardoso's first term was marked by a shock similar to that of minister Cavallo's plan, in the neighboring country. The exchange rate anchor policy was analogous to Argentina's currency board system, accompanied by a rise in interest that swept the national industrial park from subsidies and direct investments. The rule for the companies was to modernize or perish. The trade opening associated with exchange rate appreciation, however, would provide the conditions to eradicate any shortage crisis, which could be promptly answered with imports. The government started to raise funds through the successive issuing of public debt bonds, which made the public debt rise from 30% to 50% of the NDP between 1994 and 2002. In order to roll over the debts and manage interests, they adopted a privatization policy of everything that was not seen as absolutely strategic for state functions—especially those sectors linked to the energy matrix, the steel industry, the aeronautical sector, telecommunications, and mining—besides opening the capital of the largest state-owned company, Petrobras. The typical neoliberal argument was always the same: the industry becomes globally competitive, under the efficiency principle, or it has no more reason to exist in the new conditions of the global market. The remaining neo-developmentalist forces tried to politically react. The march to Brasilia, in May 1996, carried out by an alliance between the industrial barony and Brazil's main central union (CUT) became a milestone for their final defeat, as nothing resulted from the "good employers—good workers" representation. The demands of the productivist coalition on behalf of the industrial sector fell on market ears, and could not force a change of course. In this modernization "from above" sense, the neoliberal program in Brazil acted like Hegel's famous "broom of God."[46] It prompted a revolution internal to capital in order to restructure itself in the face of change and dissolve every social form that was inadequate to the restructuring of productivity and efficiency. The problem is that those who were swept were not exactly removed from their position of power in the oligarchic-corporate structure. Instead, they were re-embraced under the intact sword of the biopower bloc.

A sympathetic observer could say that, all in all, Cardoso's government program intended to pull the country out of post-dictatorship stagnation and finally connect the economy to the most dynamical globalization flows. This observer could also claim that, in this regard, the plan actually worked. According to the interdependency paradigm, the shortage of capital in Southern economies should be compensated by the activation of an internal boost of creativity, which would forcibly take care of reinstituting competitiveness and, in the long run, force a modernization from within. Yet, we must reply to this imaginary believer that creativity without antagonistic forces is not enough to transform power relations from below and reinvent the economy in other terms. The alliance with those directly linked to major properties led the Cardoso government to brutally repress the landless movement, very active throughout the decade, and trigger military forces to contain the industrial labor struggles, starting with the repression of the oil worker's strike in May 1995.[47] One fact of great repercussion gives the dimension of how the new practices of neoliberal governmentality have coalesced without disabling the biopower bloc: the massacre of Eldorado dos Carajás, in 1996, when nineteen members of the landless movement were killed by the military police of Pará State (Amazonia region) during the eviction of an occupation with fifteen hundred rural workers, in the North of Brazil. In this sense, neoliberal modernization was nothing but an internal revolution of biopower,[48] an *aggiornamento* that befitted the passage from post-colonial dependency to financialized interdependency—that is, from imperialism to Hardt and Negri's Empire.

If, on the one hand, Cardoso's government promoted a forced economy boost, opening it to the flows of post-Fordism, on the other, it ensured the repositioning of the existing power organization and its agents. At the local level, it did so by keeping most of the old colonial practices of power untouched. At the same time, on macro-level terms, it compensated the oligarchs' capital losses with extremely high interest rates, thus keeping safe the social and political blockade of biopower. Among the results was the conversion of public debt into a cash transfer mechanism, since part of the government revenue served only to repay the capital divested from uncompetitive and inefficient activities, which should be swept away if we were to experience the ideal and pure logic of neoliberal theory.

Nevertheless, one cannot level Cardoso's liberalism to the mere reduction of the State's role, since all reforms of the period—fiscal, financial, commercial, and social security reforms, as well as privatizations—happened by the state's firm hand, according to a governmental project that brought together a loose consensus of political forces. This happened, for instance, by leading the public investment bank (BNDES) to generously finance privatizations. It also instituted a recover policy for some strategic companies, which the government chose for "state reasons" among all those doomed to economic and financial infeasibility. However, the method differed completely from that of the military regime, which centrally planned everything. Instead of advancing an interventionist technocracy to accomplish the strategic planning and the industrial policy in the long term, the point now was to make very specific interventions while promoting a general reorganization of the microeconomic background, fostering the entrepreneurial society on behalf of the multiplier and demonstrative effects of globalization and increased microeconomic competitiveness.

Moreover, although following the neoliberal logic according to which the state control is an obstacle to the economy, a nest to all sorts of privilege and patronage, Cardoso's government has created a relatively autonomous space shielded from the political arena, which was related to the maintenance of macroeconomic stability. The core of the new neoliberal state form became circumscribed by activities connected to post-Fordist regulation *par excellence*: control of interest rates, exchange rates, inflation targets, and the generation of primary surplus to manage public debt. The remaining activities should be subjected to meticulous scrutiny, under the general guideline that prescribed the elimination of any inefficient or non-essential bureaucratic element from the public administration. In this scope, Cardoso created the "Ministry of State Reform," headed by Luiz Bresser-Pereira, which would participate in the developmentalist revival as a key policymaker during the Workers' Party years in government.[49]

The neoliberalism of the 1990s was an ambiguous attempt to integrate the post-Fordist global reality "from above," by way of an inter-classist alliance and a diffuse support linked to economic reorganization. But it lacked any virtuous relation with the movements and the class composition that would be necessary for creating new and open institutions. In the South, unlike the North, there was no organized Fordist society with mass consumption and a welfare state. Here, the welfare policy covered a tiny portion of the population encompassed by previous national-developmentalist gains, from the populist cycle (1930–1964) and the dictatorship period (1964–1985)—that is to say a thin middle layer and the mass workers from highly industrialized centers of the South-Southeast. This is why the breakdown of productive oligopolies and the maze of privilege associated with it allowed, to some extent, the spread of services previously inaccessible to most, being mobile telephony and higher education the most noteworthy of them. The boosting of the market and the access to those services led to an increase in income and consumption, which reversed to greater state revenue. At the same time, contrary to a "by-the-book neoliberalism," the tax burden increased from 25% to 32% of the NDP between 1995 and 2002.

Among the reforms, currency was the field par excellence of neoliberalist "biopolitical adjustments" in Latin America.[50] With the downfall of Fordism-Keynesianism consensus in the 1970s, money itself underwent a change in its very nature. Instead of being backed by the national sovereignty principle, it was now linked to the liquidity of the global financial system. The metaphor of liquefaction is adequate to indicate how a currency previously backed by the effigy of the sovereign is now "unbacked" and exists in the form of fast flows of globalization searching everywhere for high revenues or for security. This reconfiguration of the role of currency in the center of political and economic strategies is related to the turn forward to the new economy, guided by the new approach to labor, with its network, cognitive, affective, and cares features.[51] The process of capital needed to learn how to occupy the productivity networks without being able to measure them according to the Ricardian theory of value, since all life as well as the entire circuit of value (production, circulation, consumption) were now invested in productivity. The classical political economists conceived currency as a neutral medium to facilitate exchange and objectively measure wealth through working time. Keynesians, in turn, pointed to its "added" role of promoting investment and consumption as the main inducing tool, operated by the nation-state, to deal with the future, irrigating propensities and expectations of economic agents with optimism.

In the neoliberal revolution, after the Bretton Woods disarticulation and the financialization of the world, currency starts to operate according to another regime, one entirely different from those proper to liberalism and Keynesianism (we develop the analysis of money and finances in chapter 5). At first, in fact, capitalism struggled to restructure itself, resulting in the successive crises from the late 1970s throughout the 1980s around debt and inflationary pressures, signs of a broader and deeper crisis. Symptomatic of a systemic imbalance, inflation expressed, above all, the confusion of the financial system when it came to containing and exploring new flows of social productivity under post-Fordist conditions. There began a long period of pragmatic comings and goings, of insurmountable tensions, always on the verge of the dreaded day the system would crack, a fear that punctuated the history of Latin American countries around the big debt crisis of 1982.[52]

In Brazil, the Plano Real, headed by Cardoso, broke the vicious circle of hyperinflation. From this point on, currency control was incorporated as a permanent institution of governance, operated by the troika formed by the federal government, the central bank with actual autonomy, and the special Monetary Policy Committee. The so-called macroeconomic tripod—inflation targeting, exchange rate intervention, and primary surplus generation—became the cornerstone of economic and monetary policies from that point on.

The monetary integration, however, was not enough to restore the lost structure of national-developmentalist projects, leaving the biopower bloc more exposed to the action of movements and struggles once again. Between post-Fordist modernization and backing of old oligarchies, the founding contradictions of real neoliberalism led to another round of successive crises, starting in Asia, with the devaluation of the Thai baht (1997), followed by Russia (1998). The abrupt depreciation of exchange rates in 1999 paralyzed any positive tendency of the reforms, while the debt crises exterminated the budget surplus during Cardoso's second term, which was too weak to carry out significant social policies and provide incentives for more dynamic sectors. If, on the one hand, real neoliberalism meant an integration "from above," which made currency the basis of any viable democratic project in the future, on the other, it led this same currency to the heart of political disputes, to the possibility of turning the tables—that is, making democracy the basis of currency. Currency is now cut by antagonisms, grounds for diffuse resistance, embedded in social productivity. Hence the recognition—grudging on the part of the Left that missed the national developmentalism and the Fordist-Keynesian broader framework in which it was inscribed—that the constitution of currency as biopolitical grounds unveiled new horizons for struggles and resistances, a battle field that would be occupied by progressive governments in the next decade.

EXCURSUS: LATIN AMERICAN "OMNIVOROUS" LEFT

Oliver Stone's *South of the Border* (2009) portrays the quintessential narrative of the Latin American progressive cycle. The documentary tells the story of the coming to power of Hugo Chávez in Venezuela, the first to rise from a new harvest of pink leaders that distanced themselves from the monochromatic post-communist neoliberal order. Surrounded by overwhelming support from popular classes and the nationalist Left, Chavez in Venezuela, Morales in Bolivia, Correa in Ecuador, the Kirchners in Argentina, and Lula in Brazil openly confronted the elites, the press monopolies, and the right-wing "destituent" power. In doing so, they detached themselves from the governmentality that intensified inequality and poverty in the neoliberal 1990s. In a sense, this film was shot as a reverse road movie: instead of filming the popular protests throughout different territories, Stone walks around presidential palaces, embracing the epic discourse of Chiefs of State. *South of the Border* even goes as far as citing the fall of the Berlin Wall, claiming the new Latin American political cycle erupted against the triumphalist post-historical ideology of the Washington Consensus. In this narrative for export of the Latin American left-leaning regimes we can see the socialist Left's nostalgia for the Cold War and its eagerness to find an Outside of capital's global hegemony, even the survival of the sympathy for Stalinism.

In 2004, Mario Vargas Llosa published a review[53] of *El regreso del idiota*, an acid critique of the Latin American Left that came to power in the first decade of the new century. The book Llosa reviewed denounced the anachronistic restoration of populist, statist, and nationalist practices. The political regression happened on behalf of an ideological opposition to the waves of neoliberalism commanded by Yankee imperialism and the global financial market. From this critique, it is interesting to point out the distinction between the "vegetarian" and the "carnivorous" Left as described by the Peruvian writer. The carnivorous Left would participate in a sort of evil axis of Latin America formed by presidents Hugo Chavez of Venezuela; Evo Morales of Bolivia; Rafael Correa of Ecuador; and, of course, Fidel Castro of old Cuba. The vegetarian ones, in turn, would be composed of a more pragmatic and conciliatory group formed by the Brazilian President Lula; Nestor Kirchner of Argentina, Tabaré Vázquez of Urugay; and Oscar Arias of Costa Rica. According to Llosa, despite the occasional fiery rhetoric, the vegetarians would bring a positive trend of renewal to the Latin American Left, without having to resort to the raptures of an outdated national-popular imaginary. The carnivorous, in turn, would suffer from the myopia of a late and reactionary socialism, illegitimately trying to dust off the Manichean mentality of the Cold War. Actually, even groups in the other side of the political-ideological spectrum share the argument of two Lefts that would be divided by degree of radicalism into red Bolivarians—Venezuela, Bolivia, Ecuador—and pink reformists—Brazil, Argentina, Uruguay, Paraguay, Nicaragua, El Salvador, Chile. Also according to analyses "further to the Left," the carnivorous would be arranged around a new socialist horizon for the 21st century, while the others would be no more than reformist coalition governments, more aligned to the 20th-century social democracy, in pursuit of a better social regulation of capitalism, echoing the Fordism-Keynesianism imaginary from last century.

However, the contour lines between Latin American red and pink leaders are much more complex. The variations among them are not classifiable according to a simple litmus test. There is no sense in a dichotomous replacement typical of the 1970s, subdividing the Left into socialists and social democrats according to reform or revolution categories. Manifold facts suggest a nuanced classification. The same Evo administration that nationalized the exploration of oil and oil products—converted into strategic assets for the project of overcoming poverty and for the technical progress of the industrial sector—also reproduced the macroeconomic armor prescribed by monetarist orthodoxy, to the point that economist Gonzalez Chavez speaks of the paradoxical figure of the "revolutionary neoliberal *à outrance*."[54] If Evo, on the one hand, customarily adopts the anti-imperialist rhetoric on behalf of the

Southern peoples—for instance, when defending the coca industry, whose leaf was declared sacred based on indigenous cosmovisions, on the other, he does not fail to meet the prescriptions of international organizations and Western risk rating agencies, maintaining robust reserves of foreign currency, cutting public spending, and controlling inflation through a targeting policy. Thanks to the fiscal and monetary discipline to the taste of orthodox economists, "both the monetary fund and the World Bank, in recent reports, praised what they called Mr. Morales's 'prudent' macroeconomic policies. Fitch Ratings, a major credit rating agency, cited his 'prudent fiscal management,'" to which the Bolivian minister of economy said: "We are showing the entire world that you can have socialist policies with macroeconomic equilibrium."[55]

Regarding Rafael Correa's Ecuador, the claim for a 21st-century socialism coexists well with a techno-populism, as Carlos de la Torre called it. Correa's main reference is neither the saga of the Cuban mountains nor the ferocious speeches of Hugo Chavez, but the modern South Korean developmentalist path. The "Korean utopia of the Andes," as the Argentinean essayist Pablo Stefanoni named it, materialized in the Ciudad del Conocimiento Yachay or in the FLOK Society, the oasis of the digital culture which attracts artisans of free software, the knowledge economy, and hackers' ethics.[56] In Brazil, Lula—once a fierce unionist militant, when his party was the opposition during Cardoso's government—not only fully extended the macroeconomic scheme of his predecessor, naming an economist linked to that government, Henrique Meirelles (who would become the minister of economy in the Temer administration, after the impeachment of Dilma Rousseff, in 2016), as president of the Central Bank of Brazil, but also furthered the fiscal adjustment measures, concluding the reform that reduced social security rights in 2003. Six years later, after Brazil had been often promoted to a high standard for capital flow by risk agencies, Obama even said Lula was "the man" during a G20 meeting.

The differences between "carnivorous" and "vegetarian" are also debatable when comparing South American progressive governments among themselves. Nestor and Cristina Kirchner are famous for embracing the agenda of the minorities, human rights, and abortion legalization, a list of triumphs for the Left; in turn, in Bolivia and Ecuador, the reproductive rights and LGBT issues have been subject to setbacks and reluctances, with Rafael Correa speaking against a "pro-abortion agenda" and talking about "excesses in gender ideology." Meanwhile, in Uruguay, abortion and marijuana were legalized with President Mujica; when it comes to Lula and Dilma Rousseff (both from the Workers' Party) in Brazil, notwithstanding a combined total of 13 years as head of government with approval rates that peaked at 80% in some periods, abortion is still prohibited and the legalization of illicit drugs was never even considered on the agenda. It is important to emphasize that these two prohibitions have terrible biopolitical consequences in terms of clandestine practices of abortion and mostly of the "war on drugs" in favelas slums. Instead of a public health problem, drugs are still approached by the authorities under Nixon's paradigm of the "war on drugs," whose combative engagements are one of the major causes of urban violence. The "war on drugs" in Latin America induces police corruption and brutality and fosters a billionaire economy that involves illegal production, the disposable work of black and poor people—killed by the dozens of thousands per year—and a deep-rooted financial network of money laundering. It also became one of the most resilient power strategies to control the poor and manage the tensions elicited by poverty. In fact, there is no lack of state in poor areas controlled by organized groups of drug dealers and/or violent paramilitary vigilantist militias;[57] it exists under a different structural condition, something we could call a molecular *state effect*, in which the governmentality proper to Southern biopower operates. The "war on drugs" declared by Nixon in 1971 is part of the neoliberal power method to handle the unbearable pressures brought by the circuits of valorization and exploitation of the poor that Lula and Dilma Rousseff maintained along with their 13 years of Brazilian government.

In Bolivia, Evo is compared to the mythical figure of Tupac Katari, an Aymará rebel leader of the 18th century, but the neo-developmentalist project implemented by his government systematically circumvents the demands of the indigenous movements, sometimes invading their territories. The ultimate example of this was the construction of roads over the Territorio Indígena y Parque Nacional Isiboro-Secure (TIPNIS). In Brazil, the situation is even more critical. Here, old developmentalist projects from the military dictatorship were executed, multiplying the construction of mega-dams by large contractors, affecting whole ecosystems in areas populated by indigenous people and riverine fishermen. In fact, the energy and infrastructure mega-project for the Amazonian hydro-mineral complex is similar to the Second National Development Plan (II PND), which was left unconcluded by the General Geisel government (1976–1981). This reveals a disquieting (and authoritarian) consensus between Right- and Left-wing political forces of the 20th century when it comes to development projects. Remember that, in late 1970s, while Argentinean and Chilean military regimes were applying "pure blood" neoliberal recipes, Brazilian dictators were in the countercurrent, adopting developmentalist policies based on the positivistic practices dating back to the destruction of Canudos by the National Army.[58] The traditional indigenous peoples in Brazil are not even considered as a variable in the energy and infrastructure federal programs, and have been openly treated by influential government leaders as an anachronistic impediment to major projects and natural resource exploration. There is also the government's explicit assumption that "history 1," as described by Chakrabarty, would be the inescapable national course. From their perspective, those traditional peoples would be a pre-capitalist obstacle to the arrival of the "civilizing progress" and to the assertion of national sovereignty—often socialist in tone, not to mention that the indigenous movement would be, according to that same perspective, manipulated by obscure NGOs at the service of Anglo-Saxon imperialist interests in the Amazon.[59]

As for the anti-imperialism so present among the socialist Left of the 1970s, we just have to consider Venezuela, the supposedly most "carnivorous" (most red) among the progressive governments. The largest recipient of Venezuelan considerably profitable exports, mainly linked to the sale of hydrocarbons, is the United States. Ecuador, although Llosa has included it in the "axis of evil," has nothing less than a fully dollarized economy, similar to the Argentinean case in the 1990s, during the neoliberal government of Carlos Menem. In addition, Correa's government controls oil exploration in the region with a solid partnership with major multinationals, such as Chevron Corp and Texaco Company, not to mention the Chinese state enterprises that act very much like any other company in the world market. And the examples of this new interdependency are manifold: the Chinese government has joined Bolivia to launch the first satellite of the Andean country, symptomatically called "Tupac Katari"; it has built a military base in Argentinian Patagonia, which involves a treaty guaranteeing territorial cession; and is now planning one of the largest civil construction projects ever, a new interoceanic canal, to compete with the Panama enterprise, which crosses Nicaragua from east to west. The mega-project, run by Chinese communists and Nicaraguan Sandinistas, has been treating indigenous and peasant communities along the way as little flowers to be trampled on in the march of progress. In Ecuador, the banners reading "Away with China" first appeared, in a decisive way, during the mass protests of May and June 2015.[60]

Another debate that the "carnivorous" red label is unable to express concerns the relationship between the progressive government and the indigenous movements that had joined the fight against neoliberalism in a previous moment. To what extent the indigenous class counter-power is integrated with the "red" governments, could it determine a qualitative change in governmentality? Although, in the media, they bear symbols and referents of indigenous peoples, the governments of Bolivia and Ecuador have repeatedly relegated their role to the background, almost in a multicultural inclusion of features and symbols. The constitutional claim for the indigenization of power—and not just for indigenous people in power—did little more than to highlight identity factors and cultural colorful elements. In this sense, both governments seem to deal with the matter as a subsumption project for accommodating differences into "history 1" and nation-state political unity instead of facing this counter-power as a perspectivist force capable of decolonizing forms of biopower. The plurinationality of the power matrix, formally included in the new Bolivian and Ecuadorian constitutions, has been neutralized in terms of its transformative power, becoming a mere frame for conflict management and political representation.

Thus, cosmopolitical practices are reduced to "pachamamic" cultural attributes, now cynically used in order to patch the cracks of governance by invoking the ethnical identity between the leader and the people. Bolivian Vice President Alvaro Linera theorized about the notion of "Andean-Amazonian capitalism"[61] as a new creative force for an alternate development, one shaped by Aymara, Quechuan, and Guarani cosmovisions. He and his political intellectual group found inspiration in the critical lecture of René Zavaleta who, at the same time has said that "Bolivia will be Indian or not to be" but "the heterogeneous social composition (because of the indigenous presence) is a limit for the emancipation struggle.[62] However, in recent years, Linera has been the leading intellectual and political voice of the neo-developmentalist model. He emulates a new version of the TINA ("There is no alternative") principle. Linera and other national socialists now present the once neoliberal rhetoric as an unavoidable strategic project for the progress of the small Andean country. According to him, before globalization became a reality, it would be inevitable to resort to extractivism in some measure, even against indigenous movements, in order to remedy poverty, implement education policies, and counteract the old entrenched Bolivian elites, eager to retake power and annul all progress.

In the case of Ecuador, noteworthy is the closing of the Pachamama Foundation by Correa, his vigilantism tendencies regarding the Internet, and the bitter split between government and the main indigenous movements, which symbolically withdrew the command baton given to him in 2007.[63] Although presidents Rafael and Evo present their respective governments in international forums as being traversed by the eco-community ethics of *buen vivir* (literally "good living," or Sumak Kawsay, in Quechua, and Suma Qamaña, in Aymara), their predominant governmental projects are solidly founded on a modernizing national developmentalism in which the "good living" does not essentially differ from the old liberal and meritocratic model of inclusion through work. Sometimes, it is even considered an awkward barrier between old archaic Ecuador and the promised future.[64] According to Pablo Stefanoni, instead of the thesis of the two Lefts described by Mario Vargas Llosa, it is possible to identify a common thread—all distinctions and singularities of their trajectories being considered—among progressive governments. This is what he calls "lulization"—that is, the Lula government as a possible transformation tendency, toward a neo-developmental model.[65] At the turn of the 2010s, the apparent political, social, and economic success of the Brazilian neo-developmentalist project during the PT (Workers' Party) administrations in Brazil would better define something like an epochal change and a distinguishing mark of a new era. We will resort to Stefanoni's synthetic definition of "lulization" in order to trace a genealogy of continuities and discontinuities between the 1990s real neoliberal night and the so-called "post-neoliberal" stage in the Pink Tide in Latin America politics. Less than a direct break with neoliberalism, this progressive cycle meant the convergence to an agreed and gradual movement toward change, like crossing the river by groping the stones. For the Argentinian author, "lulization" was the institutional and historical synthesis of the previous insurgent cycle, the moment when Latin American governments reached a point of maturation. "Lulization" blurred the line between carnivorous/reds and vegetarian/pinks. Around 2010, even the most emphatic discourses pro revolution and anti-imperialist struggle were toned down to more pragmatic and reformist promises. Faced with real problems of internal conflicts and recognizing the unsurpassable horizon of neoliberal globalization polarized by the US-China seesaw, the progressives had to admit a certain degree of political realism, leaving the red flags for militancy or just forgiving it. Within this context, progressive governments deployed two strategies mutually articulated: economic development and social inclusion. This mathematical operation at the heart of this trend—growth plus policies for the poorest—was considered the most precious achievement of progressive governments, which would be at risk in the case of neoliberal restoration. Lulization of reds and pinks meant falling in love with the possible. Its main objectives were keeping the pace with "development with a social face" and preventing the return of neoliberal defeated political forces. In this regard, the recurring criticism related to economic reprimarization, destructive extractivism, as well as the disrespect for minority and environmental rights during the implementation of development projects, are sent to a second plane of "fine tuning." The route could be corrected without altering the main social-developmentalist pact in order to avoid any setback. Thus, there is this criticism from inside the movement that recognizes the environment and minority rights as inescapable issues, but claim no more than adjustments of course, the correction of occasional abuses and localized excesses. This ends up *ratifying* the hard core of the neo-developmentalist project, the backbone of progressive governments.

According to the lulization approach, Lula is, of course, the prime example of a leadership that would have won thanks to the ability of articulation, negotiation and composition around the program of national economic growth with social gains for the poor. Indeed, President Lula (2003–2010) was careful to remain legitimate while talking to business people in the morning, unions and manufacturers in the afternoon, and social movements and students in the evening, without being constantly questioned about contradictions and incompatibilities. Besides, he was successful in keeping himself equidistant between two key positions: he was the leader of the poor, one of their own, capable of asserting himself among the bigoted elites and also a lobbyist for the major "national champion" companies (the Global

Players) in their projects in and outside the country. Lulization forged an apparent win-win equation around developmentalism, and Lula never stopped declaring that both the worker and the entrepreneur could win. He just forgot this tiny detail: there are no rabbits in Australia.

FROM LULISM TO SAVAGE LULISM: THE END OF THE LEFT-LEANING REGIMES IN LATIN AMERICAN

Let's start at the end. We write these pages in September 2016. President Dilma Rousseff has just been removed from power after a sequence of investigations that point to campaign slush funding and "creative accounting" (another way to say she broke fiscal and budget laws), not to mention two years of constant mobilizations both in the streets and social networks, which peaked in March 2016 with millions of protestors throughout the country; a crushing parliamentary defeat with more than 2/3 of both representative houses voting in favor of the impeachment; and, finally, the homologation of the impeachment process by the supreme court. The 13-year grip of PT on the government ended amidst one of the worst economic crises ever seen in the country: the GDP fell more than 10% over the last two years, the unemployment rate doubled to 11% of the economically active population, and the labor share of national income drastically plummeted. At the same time, multiple judicial investigations—of which Operation "Car Wash" (Lava Jato) is the most spectacular one—are uncovering a billionaire scandal at the core of the PT's coalition government and the political caste in power involving graft practices, kickbacks and bribery in general. Even Lula is implied is some of those investigations, whose aftereffects seem to be the last nails in the coffin of the party's future aspirations. In neighboring Argentina, the candidate for succeeding Kirchnerism lost the presidential election to Mauricio Macri in December 2015. Just as Brazil, the country was also in the middle of a deep recession and corruption scandals already involving, former president Cristina Kirchner. Besides, the Venezuelan government of Nicolás Maduro, Chávez's successor since 2013, seemed to be on the brink of a dramatic and irreversible downfall, as basic goods shortages, hyperinflation, failure of public services, and widespread popular unrest have made the country ungovernable.

Last year (2015) was the *annus horribilis* of the Pink Tide in Latin American politics. This was the year when most progressive governments were defeated on their own terms—that is, by choice of the majority of voters. Governments that have so often invoked popular vote have lost the support of the poor. In the last and lost election, Kirchnerism could only offer as presidential nominee a candidate whose political career was shaped during Menemismo, suggesting an inner exhaustion. In Venezuela, the opposition has beaten the *Chavista* party by 16% of votes in the National Assembly elections, placing the legislative power in a collision course with the executive branch of government. In Ecuador and Bolivia, the winds of change manifested themselves in the local elections. Soledad Chapetón, young opponent of Evo Morales, was elected mayor of El Alto, second biggest city in Bolivia and the well-known plebeian heart and soul of the Gas Wars of 2003, pointing to an internal erosion of supporters. The same happened in Ecuador, as Mauricio Rodas, young politician of the new opposition party, got elected in Quito, the country's capital, right after June 2015, when the mass uprisings led to the intensification of the political crisis. The election of his allied candidate in 2017 (Lenín Moreno) did not prevent the conflict between Correa and the new President and this because of the impact of corruption investigations coming to Ecuador from "Car Wash" Brazilian investigations.[66] Finally, 2015 was the year that anticorruption protests sparked in hundreds of Brazilian cities, with an increasing focus on removing Dilma, Lula, and the Worker Party from power. In March 2015, government's popularity dropped to a record single-digit percentage, the lowest in history, even lower than the annual inflation rate.

Still in 2015, the analysis on the exhaustion of the Pink Tide starts to take shape[67] and, as of 2016, it is decisively consolidated as a central issue. What happened? What remains from the Pink Tide? How to evaluate change? First of all, it is important to avoid drawing a line between before and after in terms of a progression versus regression dichotomy. This would imply homogenizing the cycle according to a linear narrative that goes forward or backward depending on the party in power. There were multiple lines of continuity and discontinuity, a dynamic configuration of tendencies. A dualist approach would also mean losing the cutting edge of criticism, the opportunity to put into question the "progress" of progressivism (and the "regression" of regressivism as well). Perhaps we should be more precise and interrogate *what* continues and discontinues, *which* progress, *for whom*, and *how* it works at the governmentality level. The dynamic mosaic of tendencies and tensions of the last decades cannot be reduced to a unified political will, held by this or that political force, with a clear distinction between before and after. This *reductio ad* political risks the flattening of the entire problematic field, where collectives and movements strive to effectively entangle their strategic action, thus disarming the possibility of intervention and thought.[68]

There is a second analytical flattening, related to focusing exclusively on the macropolitical process, which disregards the way governmentality operates; that is to say it neglects the workings of "neoliberalism from below." Micropolitics cuts across the otherwise simplistic boundaries of "regression" and "progression." This other methodological insufficiency, so to speak, dismisses practices and logics that go beyond the political system. These practices and logics condition and are also conditioned by it. This reduction to the political sphere usually reverberates the common periodization that splits the "progressive cycle" in two distinct periods: first, an insurgent phase and, then, a properly institutional one. In sum, the first constitutes the moment of struggles like the Venezuelan insurrection of the Caracazo (1989), the *que-se-vayan-todos* metropolitan protests with *piqueteros* and *cacerolazos* in Argentina (1998–2001), the demonstrations that culminated in the *rebelión de los forajidos* in Ecuador (2005), and the plebeian Water and Gas conflicts in Bolivia (2000 and 2003, respectively), all of them inscribed in the defining wave of the global anti-neoliberal cycle of struggles. In a second moment, an "alternative power" jumps into the scene as a necessary step to guarantee electoral victory and government occupation.[69] It is as if there were a gap that had to be filled by political forces or else the entire struggle would lose potency due to restoration powers. Last but not least, there are other implications of this approach such as the fact that social struggles only happen against neo-liberal governments and politics: the message is that one cannot struggle against "progressive" politics. Within this stageist politicist framing, the *progressive turn* in Latin America would never be concluded unless it went through the second stage of the cycle, and that could only be accomplished if the constitutive powers were translated into a project of state occupation. This was historically attained by Lulism in Brazil, Kirchnerism in Argentina, *Chavismo* in Venezuela, Evism in Bolivia, and Correism in Ecuador, all of them symptomatically called by each leader's name. In addition to the transcendent postulate that separates constitutive and constituted forces, drawing a line between movements' *potencia* and state *power,* as well as the inadequate perspective of "good government,"[70] which considers us counselors of the prince, this interpretative method of the progressive cycle also reveals a terrible tendency to presume the primacy of strategical and political actions over struggles and demands once the constitutive and wild moment is gone. At the same time, it fits like a glove for the dialectical justification of some sort of battle of the progressives against the shadow of regression, which is often mobilized to accuse struggles and dissidences of treason and historical irresponsibility. In short, this approach gave rise to a bloc of unconditional supporters of the progressive governments—*governismo*, in Portuguese, *or oficialsmo,* in Spanish—in the name of long-term hegemony. Evoking a famous warning from Foucault to militants, we could say this is a kind of "falling in love with power" which eventually turns people into its functionaries.

According to the anthropologist Salvador Schavelzon, the end of the pink tide cannot be read as a "reflux" as well, as if it meant a mere reaction from the elites that, from the Outside, would have stormed the Kremlin. Nothing could create a more mystifying cloud around the issue than erasing the multiple imbrications between progressives and conservatives or leftists and rightists within the government real practices. For him, in the 2010s, the contradictions reached such a point that the progressive narrative became untenable and even dogmatic. Maintaining state power has become an end in itself, reducing the political scope to furthering the endless alliances with oligarchic, elitist, and conservative sectors, which originally were the very categories that should be confronted from a genuine leftist perspective. Schavelzon draws attention to the inner reasons of the downfall, especially the ones regarding the neo-developmentalist matrix, based on extractivism and the major national companies' monopolistic policy. Besides overburdening all governmental negotiations and projects with corruption, whose levels have jumped to the scale of billions, this approach to development and industrial policies was not able, in any case, to achieve the original developmentalist goals—that is, foster industrialization and eliminate the dependency from commodity export and its unfavorable terms of trade. Worse, progressive neo-developmentalism led to a serious side effect, meaning massive rights violation, environmental destruction, and the systematic oppression of struggles on behalf of the "big strategy." All dissidence would be punished, as it was seen as automatically naive, irresponsible, or outright treason, for it was "playing the game of the Right." For Schavelzon, although the left-leaning regime years brought about some relief for the poorest, better income distribution, and significant social policies (in education, for example), the overall balance was eventually negative regarding structural changes in "inequality and the economic matrix." "Concerning the geopolitical sphere," he adds, "the increased repression and criminalization of dissents articulates with the left-leaning governments view of the East, coming closer to authoritarian regimes like the ones in Russia and China—by now, also devoid of any anticapitalist and emancipating horizon—both in discourse and economic approach."[71] In 2013, seeking to apprehend a more intricate dynamics of social mutations during the governments hegemonization by the Workers' Party in Brazil, we have theorized on the outbreak of *Savage Lulism*,[72] which had a striking expression in the metropolitan uprisings of that same year. Following the world series of protests which started in Egypt and Turkey, that was the *annus mirabilis* for the struggles in Brazil. On that occasion, approximately 30 Brazilian Municipal Chambers or State Parliaments (Assembleias Legislativas) were occupied by demonstrators; the streets and squares of more than 400 cities were covered with Baroque masses of protesters (the *ugly June journeys*); a consistent campaign against institutional racism, state violence, and corruption was capable of building up a strong voice in national and international media; and an all-time record of more than 2,000 strikes were registered nationwide.[73] In that scenario, we sought to decode the vertiginous succession of facts in order to point to the class composition that, according to our assessment, was at the brain and heart of those struggles.

Some questions were unavoidable. How come the biggest uprising of 21st-century Latin America was happening right in the country whose "Lulist drift," combining economic growth and social inclusion, was a successful reference for the entire belt of progressive governments? What happened to the subjectivities? This line of questioning does not imply we were trying to determine a linear progression from the emergence of the "new middle class" to the rise of a class antagonist power. Instead, the method started by identifying several singular arrangements—among government, subjectivities, and capitalist transformations—that, in 2013, ended up converging in a multitudinous movement without unifying organization centers.[74] In order to do so, it was important to first understand Lulism "from below," the assemblages that constituted it and lines of flight from its more institutional elements.

The arrival of the Workers' Party (PT) at the Brazilian federal government in 2003 was preceded by a series of conciliations whose goal was gathering support from wider economic and political forces to the campaign and party's historical leader. That was Lula's fourth attempt at a presidential election, having lost in 1989 to Fernando Collor, who was later impeached, and in 1994 and 1998 to Cardoso. The next time, the PT decided to handle things differently. The electoral marketing campaign worked around Lula's image, calling him "Lulinha paz e amor" (Lulinha peace and love). This novel approach acquired a manifesto when the candidate signed a public letter[75] marked by several compromises. Among other things, he assured property rights and macroeconomic equilibrium in neoliberal terms. Reinforcing even further this new moderate image, Lula chose José Alencar, a super-rich industrial entrepreneur and leader of an evangelical political party, as his vice president. Several militants became bitterly disappointed once Lula got elected. But the fact that his first mandate (2003–2006) did not promote a single structural change, especially in the economic area, should have come as no surprise. Actually, the first policies of the new president strictly followed his predecessor's line; he even passed the social security reform that Cardoso was politically unable to approve. Lula not only preserved, mostly untouched, the macroeconomic policy founded on the orthodox tripod—positive primary balance of payments with central bank independency, strict inflation targets, and floating exchange rates—but also significantly raised the benchmark interest rate, which reached its peak at 26.5%, the highest in the world at the time. This caused the first dissent within the PT, leading to the creation of PSOL,[76] a left-wing party formed mostly by university professors and students. The *Fome Zero* (Zero Hunger) program was the only real innovation during the first year of the PT administration. Following propositions from left-wing intellectuals of Catholic tradition, the program allowed the Ministry of Agrarian Development to implement the "universal right to food." But, as was predictable, this program did not work: the logistics of the distribution were becoming more expensive than the food. At the turn of 2003–2004, there was a qualitative leap when they reformulated the social program: instead of providing *in natura* grants such as food and offering services like professional training courses, the government concentrated exclusively on direct cash transfer policies. *Bolsa Família* was then chosen as the backbone of a new social protection system based on direct income distribution. Every single social policy was gathered around that program. It gained strength over the following years, eventually turning into the world's largest income program in terms of the number of beneficiaries and volume of funds. In December 2012, almost 45 million people were receiving the benefit which amounted to US$ 10 billion, corresponding to 0.5% of the Brazilian GDP. The enormous success of *Bolsa Família* among the poorer made it the showcase number 1 in every political (and electoral) event Lula went to, and it quickly became the flagship for him and the PT, both nationally and internationally. Despite the outbreak of a first corruption scandal in 2005, when almost the entire PT leadership was exposed for bribing dozens of parliament members from the opposition in order to gather votes, Lula had enough popularity to be re-elected in 2006. If, in 2002, the bulk of Lula's constituency was concentrated in the big, industrialized cities of the South and Southeast areas of the country, particularly within the urban middle classes and among regularly employed workers, in the 2006 campaign, the situation changed. Now, most of his voters came from the poorer and the so-called "lumpenproletariat," with a decisive majority in the North and Northeast regions. On that occasion, the right-wing parties, the corporative media, and also the left-wing opposition (PSOL among them), spoke of "populism" and even accused *Bolsa Família* of being a way to buy voters. In parallel to the revolution brought by the *Bolsa Família* program, the government promoted a gradual yet consistently regular increase in minimum wage (MW), with annual increments above inflation, during the whole time they were in power. Since MW works as an indexer, its increase triggered a cascade of readjustments in the value of other benefits, pensions, and programs, directly reaching a significant fraction of the population. Finally, completing the picture, popular credit lines were abundantly encouraged, including to the poorest percentage of the population, previously without access to this benefit. Credit expansion coincided with the expansion of the banking sector, fostering the injection of new capital into the economy through financialization. The result was the opening of millions of new accounts in public and private banks and, with greater access to credit, the possibility of buying in installments and mobilizing capital, however small it was.

The Lulist turn, headed by direct cash transfer policies, determined a deep and lasting change in behavior and social relations. For example, while the Cardoso period, in the 1990s, was referred to as the "chicken era," the Lula period was characterized as the "red meat era," when the possibility to consume red meat became a sign of social inclusion, since this used to be restricted to the higher income layers. The majority of the population finally gained access to previously exclusive venues like airports, restaurants, and beauty salons. All that, along with qualification policies for small entrepreneurs and milder legislation for small businesses, has promoted a wave of entrepreneurship "from below," which quickly spread throughout the Brazilian society, radiating circuits of income, consumption, and credit. Another government policy of great impact was the indirect financing of affordable housing through subsidies for construction materials and facilitated access to micro-credit, boosting a formerly diffuse self-building movement.[77]

As we have seen, Foucault defines neoliberalism as the art of governing which depends on an immanent principle of competition and efficiency spread throughout the social sphere, thus producing the enterprise-society. Lula's policies helped adjust the equation between capital and labor, allowing the penetration of financial capitalism under post-Fordist conditions, which already existed in Brazil. With Lula, neoliberalism appeared as what it really is: not a regime of exclusion, but of inclusion by modulation of all kinds of precarious life conditions. The neoliberal reforms and inflation control had already prepared the grounds for this transformation in scale and depth. We could say that Cardoso prepared the terrain for neoliberalism "from above," launching the institutional basis for the next step. But that was not enough for capitalist modernization, as the Asian and Russian debt crisis in the late 1990s showed us. Only Lulism in the mid-2000s was able to give substance to "neoliberalism from below," providing capital liquidity and connectivity to an otherwise truncated and chronically jammed social productivity. Just by the direct financing of the social tissue, capital could reorganize itself *from within*, weaving the molecular web of neoliberalist productive relations and its corresponding subjective figure, the *homo oeconomicus*.

To some extent, the leftist critics of Lulism were in fact correct when they diagnosed the perverse increase in debt, the financialization of life, and the replacement of the industrialization "hard" project with the spread of businesses and enterprises in the service sector. In that sense, this is really not a case of "post-neoliberalism." But most of those critics were correct in just one aspect of their analysis, which by the way coincides with the perspective of capital subsumption (*h1*). In parallel to the described process, there was also the accelerated spread of biopolitical resistances along the new lines of the expansion of neoliberalism. Throughout the 2000s, one cannot lose sight of this *emerging process,* already in full biopolitical conditions, which cannot be reduced to mere absorption to the market values, consumerism and "neoliberal ideology." This refers to a production of subjectivity "within and against" the neoliberal matrix that biopower has undertaken in the South in recent decades. Lulism cannot be restricted to a one-dimensional analysis, for it is an intricate phenomenon, following (and inducing) global capitalist transformations. What matters is the production of subjectivity.

These are the circumstances in which former spokesman of the government, the political scientist André Singer, launched the concept of "Lulism." In *Os sentidos do Lulismo*,[78] he argues that the mix of social policies with the charismatic figure of Lula led to the consolidation of what he calls the "Brazilian New Deal."[79] This pact was based mainly on the government's ability to combine economic growth with income distribution, a supposedly win-win logic that would please both capital and the workers.[80] The result was a new social composition loyal to Lula and the PT, which the author named *subproletariat*. According to Singer, the power of Lulism is in its ability to gather support from this "inorganic and disorganized" layer of society, nurtured during the socio-economic transformations of the 2000s. The *sub* in subproletariat thus has an ambiguous connotation: on the one hand, this composition is the result of a social inclusion model; on the other, it does not correspond to the class structure present in Fordist-Keynesian societies in Europe and the United States (or in socialist societies), which is to say that the subproletariat is incapable of an organic class will, which brings it nearer to the figure of the lumpenproletariat. Still, according to Singer, before coming to power, PT was mostly guided by the anti-capitalist spirit. However, after 2002 the party abandoned the class grammar and adopted a flat-out populist scheme, opposing the rich and the poor, the people and the elites. Thus, Lula represents a new stable setup for the functioning of the political system, with lasting repercussions on the electoral level. On the one hand, he functions as a direct representative of the subproletariat. Without going through party mediations, social movements, and unions, Lula allows for a connection that relies on the emotional charge he can mobilize as a leader. On the other, he acts as a pact maker and guarantor in his relationship with the business community and private banks, whose demand for higher profitability must also be met. The result of this double pincer of Lulism is a blend of conservative agreements and slow-paced reformism.[81]

To some extent, Singer reproduces the leftist critique of Lulism by conceptualizing the subproletariat as disorganized and uncritically adherent to neoliberal values. He also reproduces the traditional literature of national formation,[82] which attributes the social composition in the post-colonial South to an amorphous character, an absence of strong labor values or a "firm and safe" ideological education. According to him, the Brazilian newborn subproletariat would be interested, above all, in a scenario of orderly improvement of life, and "is not able to build its forms of organization from below."[83] It turns out that such social structure was never established in Brazil, except in a localized and incomplete manner, in some areas of intensive industrialization and urbanization in the South-Southeast (where the largest Brazilian cities such as São Paulo, Rio de Janeiro, and Belo Horizonte are located). Thus, in its fluidity without structure, the subproletariat would seem practically adrift to those theorists regarding consumerism, corporate media campaigns, or populist appeals of candidates and parties with no ideology at all. In this view, subproletariat is just like the bare life conceptualized by the Italian philosopher Giorgio Agamben.[84]

It is worth noting here a certain Eurocentric tendency (of Chakrabarty's *h1* kind) disguised as Brazilian particularity, since the always unreached model seems to be really based on the Western post-war Fordist working class. We should also point out the negative connotation given to the *sub* prefix, supposing the existence of an incipient rudimentary underclass. It is no coincidence that his analysis of Lulism ended up corroborating the very PT government, seen as the best that could be provided within the correlation of forces given by class disorganization. After all, if outside of the PT's historical process there is only an amorphous and potentially conservative social broth, it is not hard to understand the claim that the Lula government is the very limit of the achievable within the existent relation of forces between the poor and the elites (should we accept this contradiction as structural in the first place). Singer's version of Lulism largely resembles Stefanoni's idea of Lulization: a "passionate pragmatism of the possible," a social process without new political subjects, without emergences from the subjectivities. In both cases, the leading role of the experienced changes is assigned to the process of capital alone, under which all resistances are fragmented, and class dispersed into post-modern entropy. The government is then left with the sole task of regulating capital, as the lesser evil in the face of a full-fledged neoliberalism. One time more, we are in the theoretical context of the autonomy of politics.

The income, consumption, credit, and access policies provided by the Lula government, rather than fabricating subtypes of an always chimeric proletariat, as envisioned by a nostalgic Left—which, by the way, had never experienced those conditions in Latin America—contributed to a new adjustment of governmentality. Moreover, elements of subjectivity were activated and dispersed in ambivalent ways throughout the social fabric.[85] Lulism meant inserting the newly included into a social relationship with capital already in post-Fordist conditions. We can speak of subproletarianization in the Lula years, as long as we avoid the automatic negative connotation given to the *sub* prefix. We rather understand it as a full proletarianization under the given conditions of fragmentation, flexibility, and mobility. In this scenario, instead of considering the Brazilian social composition as the mere emptying of the notion of class, we should approach it from the perspective of *biopolitical resistances*—that is, new cultures of resistance, greater urban and rural mobility which can mutate into mobilizations, as well as new productive arrangements that are flexible and networked—for it allows us to identify the powerful dimension of the *sub*. In other words, we must envision the fluid strength of something which would not be plain disorganization, but rather a social and labor organization under new conditions, an *internal* subjective activation of the biopower matrix.[86] Therefore, a class analysis must accommodate both sides in an articulated manner: on the one hand, the constituted power as a process of violent exploitation, extraction of productivity, and racism in flux; on the other, the constituent power, which involves biopolitical resistance movements, strategic action of forms of life, and the re-appropriation of surplus.

There is an exit *toward the inside* of the very neoliberal governmentality, which was drafted in the breaches of Lulism. This unexpected route, which went on unnoticed to (and undesired by) most of the formulations, contributed to the production of subjectivity based on the social heterogeneity of the South. It is in this sense that we refer to this process as "savage Lulism,"[87] which does not correspond to the concepts of Lulism (Singer) or Lulization (Stefanoni). We wish to point out a *clinamen* in relation to the conservative pact and gradual reformism associated with the developmental project during Lula's years as president. The straight line of Lulism/Lulization was unable to cope with the line of savage Lulism, creating a gap that, by 2013, was overflown with new tendencies when Lulism got out of control and was definitely over as a government dynamic.[88] A new cycle was then inaugurated, as we shall see, also *against PT*.[89]

THE NEW ECONOMIC MATRIX: THE PROGRESSIVISM OF DISASTER

Right now, speaking of the Brazilian recent past is an extremely difficult task, almost inglorious. There are so many passionate mystifications and international campaigns either for or against Dilma's impeachment (or "parliamentary coup," if the reader prefers, but there is also a certain enjoyment in leveling all analyses within the same structured narratives. In the arena of debate around the PT government's downfall, since 2014, we are facing an environment saturated with moral charges and silence spirals which paradoxically interdicts thought at the exact moment thinking is most needed. We find again this authoritarian situation in the debate about the tragic social and economic collapse of chavist Venezuela Imperative demands, gregarious code words, the complot mantra, moral panics, and energy-costly "cultural wars" setting the tone of the discussion on what has led Brazil to the current conjuncture, to what has been trivially called a "perfect storm" or, another cherished image of these days, "free fall." What happened between the 2013 urban uprisings and the 2016 climactic outcome is really astonishing given the confluence of adverse circumstances and dramatic facts. What once many considered a promise of redemption, the ultimate achievement of the country's future, has been buried in the common grave that is now the long list of national frustrations. No easy answer or categorical assessment will be able to shed light on the situation at hand. There is no use in reducing all to an elite-backed imperialist coup or an ebb moment in terms of a rematch against the progressive cycle. Those are not the case, although the Right is certainly salivating.

Moreover, before going around distributing guilt, it would be necessary to question not just the "progress of progressivism" and all structural variables conditioning historical events—which we have already partially explored—but we would also have to delve into our own concept of crisis and its respective temporality. There is a communist tradition, which dates back to Marx's *Eighteenth Brumaire of Louis Bonaparte*, that considers crisis an opportunity for transforming reality, when existing contradictions become acute and favor action. Sometimes, the conflicts within particular critical moments lead to solutions that, despite their violent nature, are still viable solutions. According to this modernist notion, the highest point of the crisis is precisely the right time for the historical subject to intervene and open the present for the future. At the edge of the possible, this creates a new tempo. Crisis then informs the agents of change that the situation has arrived at a saturation point and so a new horizon should not be far away. This crisis-oriented optimism, however, must be subjected to criticism.

Commenting on recent Brazilian events, the political scientist Paulo Arantes gives the example of June 2013 protests and the breach they have imposed on the conjuncture.[90] Within that singular temporality, the Ugly June won in every aspect. It uncovered the permanent war sustained on a daily basis against the poor, which seemed flattened on the landscape before that; it was also successful in the campaigns with the very concrete demand for lower bus fares; and it compelled the state to investigate cases of police brutality, as in the famous Amarildo case.[91] Additionally, it succeeded in breaking the broad consensus concerning neo-developmentist politics (called "Brasil Maior"), whose shattered pieces the political system was unable to recompose the following years, leading to a "destituent" crisis of governance. Nevertheless, as Arantes remarks, at the height of the "destituent" process, no historical subject jumped into the scenario with a project for the new era. There was no such subject, no constitutive force to create a new future. June had no future other than being bitterly defeated by the repressive and ideological apparatus. As he saw it, this expressed the sign of our time, and why it has irreversibly departed from the old modernist Left that believed in the future. Crises should not be understood as they were anymore, as turning points in history. According to Arantes, the June protests brought a new character to the scene, the "Left with no future," which struggles both against the usual Right and the "Left with future" (progressivist/developmentalist). This nocturnal Left lives in the new and extemporaneous time of the world, where a new "political regime of waiting," with lowering expectations, determines the sensibilities. In this new time of a disenchanted world, the "Left with no future" follows a "reactive and anti-political logic that dies out and leaves no other trace than the memory of resentments, accumulated for the next explosion."[92] For a brief moment in June, that *no-future* Left was winning. Indeed, it did. But for this very reason, it had no future and was defeated. It wasn't meant to have it. Thus, the historical restoration over the following years created its own future by "upgrading the coactive mechanisms," bringing a "total peace." We could also add that, the moment the long-duration biopower, usually embedded in the landscape, came to the fore, it showed its thousand teeth.

Having Arantes's gloomy "fire alarm" in mind, we wish to present an alternative explanation for the disaster of the progressive government in Brazil. In this section, we seek for lines of flight, especially the ones related to the June Event which expressed an *anti-conjunctural* breakthrough. For that, we focus on the blast of the Lulist Pact (in political terms), which meant the simultaneous explosion of the articulation between the "neoliberalism from below" (molecular/production of subjectivity) and the "neo-developmentalism from above" (molar/state apparatus). Neoliberalism and neo-developmentalism have never occupied opposite poles, unless in some binary narratives from the "post-neoliberal" propaganda. There are rather zones of interpenetration and assemblages in constant interaction amidst them. Nevertheless, this field of coexistence among processes must be explained through its working machine, and without any functionalist trace of analysis. The composition involving neoliberalism and neo-developmentalist finally reached the boiling point in 2013, and then it wound up burning. In short, we argue that a terminal crisis was already irreversible for the PT government since as early as 2013–2014, especially when they were unable to respond to the mass protests and occupations, acting much like a "Party of the Order."[93] Above all, the party behaved as Order by reinforcing the "pacification" project of the biopower bloc against the poor, the Black and the resistances. But, first, we must to go back to the 2008 subprime global economic meltdown, when "phase 2" of Lulism was set in motion.

Back in the 2008–2009 biennium, while the countercurrent of Savage Lulism[94] was growing under the official narrative, the Lula administration entered a new phase, which came along with the escalation of the developmentalist discourse as well as those political and theoretical trends more attuned with the old Fordist and Keynesian projects of industrialization. The debate on a true national project, in the lines of the literature of national formation, was resumed. An already blunted chain of discussions, which had seemed definitely overcome after the controversy surrounding dependency theory and the neoliberal wave of the 1990s, was reopened. In fact, the political and theoretical developmentalist movement was gaining momentum since 2007, when huge exploitable oil deposits were discovered on offshore platforms in the Brazilian sea territory, the so-called Pre-Salt Layer. With a barrel of oil costing more than US$ 100, this discovery was rapidly integrated into the official discourse, treated like a goose which lays golden eggs. During Dilma's electoral campaigns, in 2010 and again in 2014, the pre-salt geological reserves were presented as a universal ready-made solution that could remedy any national illness. If we were to believe in the government propaganda, directing part of the dividends arising from the exploitation of the new oil deposits would be enough to solve all chronic deficits in the country: education, health, infrastructure, energy, and culture.[95]

In 2008, the crisis of global capitalism exploded, triggered by unsustainable leveraging of subprime lending related to mortgage debt and a number of derivative securities. The crisis, which started on Wall Street, spilled over into the world market, causing the recession of 2008–2009. During this period, the Lula government responded with a countercyclical prescription, typically Keynesian, with incentive campaigns to boost consumption and credit. The package also included the expansion of public investments in infrastructure and in the consumer goods industry, with special focus on heavy civil construction, the automotive industry and large-scale energy production. At that same time, eye-catching billionaire packages of economic recovery were injected into the Shanghai Stock Exchange by the Chinese government, in an emphatic Keynesian maneuver on the part of the Communist Party. In Brazil, despite the lukewarm year of 2009, the country had gone through the turbulence of the global economic crisis relatively unscathed. Running counter to the recession, the country resumed its growth with a "Chinese rate" of 7.5% in 2010. This occurred precisely in the decisive year when Lula had to choose a successor, since the Brazilian Constitution allows a single re-election (already held in 2006). The momentary success of the Keynesian and countercyclical measures not only excited developmentalists, but also gave the Lula government international recognition regarding his ability to promote political and economic prosperity amidst a severe international crisis. At that time, the country had managed to promote itself as a showcase for the global future, to the point of attracting international mega-events such as Rio+20 (2012), World Youth Day of the Catholic Church (2013), the World Cup (2014), and the Olympics (2016). That was the high of the good news season for the PT government in Brazil. This news, coupled with popular recognition in response to the distribution of income and the positive results of the economy, led Lula to the climax of his political career. By the end of 2010, he was covered in international awards and enjoyed stratospheric popularity ratings (above 80%). This allowed him to successfully help Dilma Rousseff, who had no previous electoral experience, run for office and win the election. Lula chose his own chief of staff, known as the "mother" of the Growth Acceleration Program (PAC)[96] to succeed him. With a technical and authoritarian profile, Dilma received the presidential sash at the peak of Lulism, when the rest of Latin America began to regard Lula's style and program as a reference, ratifying what Pablo Stefanoni called the "Lulization" of the "carnivorous" and "herbivorous" Left of the Pink Tyde. Lulism seemed to be everywhere and the investment of the National Bank (BNDES) and Brazilian Global Players through all Latin American countries was the industrial harm of this political hegemony, something that Raúl Zibecchi called Brazilian Sub-Imperialism.[97]

Seen from Brazil's perspective, the impacts of the 2007–2008 crisis of global capitalism were paradoxical. In the affluent economies of the United States, the European Union, and Japan, the financial crisis resulted in a violent attack on the remnants of the *welfare state* (of social protection), which had resisted almost thirty years of neoliberalism. This time around, it is the very biopolitical foundations of democratic constitutions from the second post-war period that are crumbling. In this sense, the crisis seems the "Brazilianization of the world." A macroeconomic standstill within the debt cycle and the budget cuts immediately make us go back to the "lost decades" that plagued Latin America in the 1980s and 1990s. Almost ironically, this becomes apparent in the opposite direction of the interest rate curves of sovereign debt—which, by 2014, already presented this trend for a few years—when comparing Brazil and affluent countries: while the first celebrates the decline of that rate to 7.5%, the president of the European Central Bank, Mario Draghi, for instance, considers the 7% interest rate of government bonds so high as to be a "disaster."[98] On the occasion of a conference at the Sorbonne, on November 6, 2013, the President of Ecuador, Rafael Correa, claimed: "Europe makes the same mistakes we Latin Americans do."[99] Following the same line, the banker Bill Rhodes states that "Europeans must learn from Brazil, which faced problems with the foreign debt in the 1980s and solved it."[100] Paul Krugman has even declared that the "Argentine lessons" were important for Europe, claiming the country's departure from neoliberal orthodoxy was an "extraordinary success."[101] Even the ideologists of financial capitalism explicitly recognize the political dimension of debts and conclude that "risk investments and default are inextricably connected." In light of this, the advice is explicit: "Escaping the US jurisdiction would seem to be a very sensible measure."[102] However, with a good dose of irony, we can always say to the European multitudes: "Welcome to the lost decades which we, Latin Americans, know from the 1980s and 1990s." Greater job insecurity, reduced democracy, expansion of poverty and civil violence are already the context of this inflection. Just remember the cause of the London Riots, in 2011—an anti-political movement from a "Left with no future," to employ Arantes's diagnosis of the new time of the world. Generally speaking, the middle-class majority in central economies is going through an intense process of fragmentation and impoverishment, the aforementioned Brazilianization. As the German theorist who presented such a concept, Ulrich Beck puts it: "The poor are becoming poorer and the middle classes are threatened with declassification, and no one can see the light at the end of the tunnel." The "Brazilianization" of the world and labor thus appears as an alternative way to speak of the increase in precarious relations.

When we shift the perspective to Brazil (and more generally to the South) at the turn of the 2000s to the 2010s, the crisis seems to point to an opposite trend, indicating the arrival of the much desired future.[103] Thus, as announced by the Dilma government since her first term in office, Brazil could finally become bigger ("Maior"), even with a "New Middle Class,"[104] and assert itself as the country which plays in South America the same role Germany plays in the European Union. This gave rise to coinciding assessments of a new sort, which brought together much of the international media while the mainstream national media also claimed Brazil was becoming the world. Be that as it may, this world from which the country seeks legitimacy is no longer the alternative it once was. Now, paradox of paradoxes, we are facing the possibility of a *Brazilianization* of Brazil. Thus, the crisis of global capitalism is introduced in the Latin American country as a potentially big trap—a trap that the large demonstrations of June 2013 have proven to be more destructive than we could have imagined.[105]

The fact is that the search for a new growth model based on neo-developmentalist prescriptions got stuck (and eventually failed) precisely because they could not mobilize values other than those linked to an exhausted market and the ruling elite, with its banks and agribusiness groups. There is a vast amount of literature on critical economics in connection with circles of economists close to the Dilma and Lula governments supporting the developmentalist measures, from the more state-centered socially oriented view to the export-led model of growth. However, it is not hard to see they were never even near breaking out of the trap in which the economic policies got jammed. This impasse involves the billionaire subsidies given to the automotive (transnational) industry, the agribusiness and major construction companies, as well as the short-lived attempts to break the infernal cycle of high interest rates. Just as Brazil is being seduced and declared an emerging country by national and international media, and even a reference of a successful development project, the country became a constant object for concern and criticism. On a 2009 cover of *The Economist*, we could see Christ the Redeemer taking off like a missile; by September 2013, there it was again on the cover of the same business magazine, but now Christ was uncontrollably accelerating toward the ground.

The re-election of President Dilma, contrary to what a superficial observer might think, was worse than the victory of Pirrus. It was a real dive into catastrophe. In October 2014, with falling popularity and surrounded by the conjoining problems of growing inflation, shortage of investment, reduction of commodity prices in the international market, and the leakage of multibillion corruption cases, Dilma's re-election campaign had to deny that the country was at the brink of recession and that some sort of fiscal adjustment was inevitable. The entire campaign was built on the idea that Dilma was not going to resort to what her two main opponents would supposedly do—that is, employ austerity measures. Once elected by the narrowest margin ever seen in the Brazilian electoral history, the government immediately admitted that things weren't good in the economic field and, what is more, nominated a neoliberal bank executive to the Ministry of Finance, announcing austerity reforms for structural adjustment. Dilma lied twice and won. The adjustment, however, was a complete failure: first, because there were not enough political forces left for the president to accomplish the programed reforms; and, second, because the economic situation had terribly deteriorated, triggering a 10% drop in the GDP and more than doubling the unemployment rate in the triennium (2014–2016). The series of scandals involving bribery, graft practices, and illegal funding exposed by Operation "Car Wash" made the PT a major symbol of the political system's bankruptcy. In a country where the political class as a whole, no matter the party, was being openly and massively rejected by the population since June 2013, all this news dropped like a nuclear bomb. Then, throughout 2015, the streets were quickly overcrowded with millions of protesters. And, very much due to the inertia of the Left and their reluctance to charge against a red-colored government, organized groups from the conservative side were able to grab the moment and take the lead in the streets, something they did not manage to do during the 2013 cycle. At that moment, it became clear that Dilma's second term (2015–2018) was effectively over without even having started. The tie-up came in March 2016, when the biggest demonstration of all time against Dilma and the PT took to the streets of hundreds of cities, impelling the impeachment process in a Congress where a frightened political class did not hesitate to deliver the president's head on a platter in order to save itself.

NOTES

1. Juan Gabriel Valdés, *Pinhochet's Economists: The Chicago School in Chile*. Cambridge: Cambridge University Press, 1995.

2. The currency board policy involves establishing an autonomous monetary authority responsible for maintaining the exchange rate fixed or within a given range in order to remain safe from the political game.

3. "In these various ways, then, the claim of development economics to stand as a separate body of economic analysis and policy derived intellectual legitimacy and nurture from the prior success and parallel features of the Keynesian Revolution." Albert O. Hirschmann. "The Rise and Decline of Development Economics," in *Essays in Trespassing: Economics to Politics and Beyond* (Cambridge: Cambridge University Press, 1991), 6–7.

4. For a synthesis of the cat-and-mouse game between constitution of class struggles and reconstitution of capitalist unity, see César Altamira, *Os marxismos do novo século*, trans. Leonora Corsini (Rio de Janeiro: Civilização Brasileira, 2008 [2006]), 70–86.

5. Boltanski and Chiapello, *The New Spirit of Capitalism* (London: Verso, 2007 [1999 French edition]).

6. See Giuseppe Cocco, "A Cidade policêntrica e o trabalho da multidão," *Lugar Comum* 10, April 2000, 61–89. See also Maurizio Lazzarato and Antonio Negri, *Trabalho imaterial: Formas de vida e produção de subjetividade*, trans. Mônica Jesus (Rio de Janeiro: DP&A, 2001). And also Alexander Galvão *et al.*, ed. *Capitalismo cognitivo: Trabalho, redes e inovação*, trans. Eliana Aguiar (Rio de Janeiro: DP&A, 2003).
7. Gilles Deleuze, "Postscript on the Societies of Control," *October* 59 (Winter 1992 [1990]): 3–7, http://www.jstor.org/stable/778828.
8. Antonio Negri, *Dalla fabbrica alla metropoli: Saggi politici* (Roma: Datanews, 2008).
9. By 1972, Deleuze and Guattari already called it "machinic surplus." Gilles Deleuze and Félix Guattari, *Anti-Oedipus: Capitalism and Schizophrenia* (Minneapolis: University of Minnesota Press, 1983 [1972]), 232–37.
10. Andrea Fumagalli and Sandro Mezzadra, eds., *Crisis in the Global Economy: Financial Markets, Social Struggles, and New Political Scenarios* (London: Semiotext(e), 2010).
11. Yann Moulier Boutang, *Cognitive Capitalism*, trans. Ed Emery (Polity Press, 2012).
12. Concerning this insistence on a class project in the Empire's new post-Fordist conditions, which changes the nature of the struggle as well as the organization and mobilization modes, Negri e Hardt remixes Spinoza's concept of multitude, the post-modern proletariat in the dawn of the 21st century. Ibid., 417–37.
13. David Harvey, *A Brief History of Neoliberalism* (New York: Oxford Press University Press, 2007), 9.
14. Sandro Mezzadra, *Nei cantieri marxiani: Il soggetto e la sua produzione* (La Talpa: Manifestolibri, 2014).
15. Pierre Dardot and Christian Laval, *The New Way of the World: On Neoliberal Society* (London: Verso, 2013), 11.
16. Naomi Klein, *No is Not Enough: Defeating the New Shock Politics*, (London: Allen Lane, 2017), 10.
17. Gilles Deleuze and Félix Guattari, "Plateau 9—1933: Micropolitics and Segmentarity," in *A Thousand Plateaus: Capitalism and Schizophrenia*, trans. Brian Massumi (London/Minneapolis: Minnesota Press, 1987 [1980]), 208–30.
18. Ibid., 213. Original emphasis.
19. Dardot and Laval, *New Way of the World*, 12.
20. Gilles Deleuze and Félix Guattari, "Plateau 13—7000 B.C.: Apparatus of Capture," in *A Thousand Plateaus*, 464–66, 470.
21. Ibid., 437.
22. Ibid., 468.
23. Deleuze and Guattari, "Plateau 13," 436. Original emphasis.
24. *The Courage*, 10.
25. Andrew Ross, *Nice Work if You Can Get It: Life and Labor in Precarious Times* (New York, New York University Press, 2009), 53–76.
26. Ibid., 74.
27. On the genealogy of neoliberalist governmentality in China, see also Aihwa Ong, "Self-fashioning Shanghainese: Dancing across Spheres of Value," in *Privatizing China: Socialism from Afar*, ed. Li Zhang and Aihwa Ong (Cornell University Press: Ithaca-London, 2008).
28. See, for instance, the chapter on China in Harvey's *A Brief History of Neoliberalism*, 120–51.
29. Veronica Gago, *La rázon neoliberal: Economias barrocas y pragmatica popular* (Madrid: Traficantes de Sueños, 2015).
30. Hernando de Soto, *El otro sendero—la revolución informal*, (Lima : Instituto Libertad y Democracia, 1986). With a prologue wrote by the Peruvian writer Mario Vargas Llosa, the book proposed a clear alternative to the leftist Maoist guerrilla group named *Sendero Luminoso*. The book was translated to English with the title *The Other Path: The Economic Answer to Terrorism*, (New Yoir: Harper & Row Publisher, 1989). Another book with a development of the same approach by Hernando de Soto is *The Mystery of Capital* (New York: Basic Books, 2000).
31. Ibid., 28.
32. Ibid., 25.

33. Ibid., 319–29.

34. Oswald de Andrade, "Manifesto antropófago," in *A utopia antropofágica*, 4th ed. (São Paulo: Globo, 2011), 67–74.

35. Francisco de Oliveira, "Crítica da razão dualista" [1972], in *Crítica da razão dualista/O ornitorrinco* (São Paulo: Boitempo, 2013), 25–119.

36. Fernando H. Cardoso and Enzo Faletto, introduction to *Dependency and Development in Latin America* (Berkeley: University of California Press, 1979).

37. We will develop this point further in chapter 5.

38. Giuseppe Cocco, *MundoBraz: O devir-Brasil do mundo e o devir-mundo do Brasil.*

39. Tim Marshall, Prisoners of Geography, (London:CPI Group), 2015, p. 253.

40. Bruno Latour, *We have never been modern* (Cambridge, MA: Harvard University Press, 1993).

41. Eduardo Viveiros de Castro, "The relative native" trans. Julia Sauma and Martin Holbraad, *Journal of Ethnographic Theory (HAU)* 8, no. 8 (2013).

42. "Authoritarianism is a turned page in the history of Brazil. However, there is a piece of our political past that still obstructs the present and hinders the advancement of society. I refer to the legacy of the Vargas Era, its autarkical development model, and its interventionist state." Fernando Henrique Cardoso, 1995. *Apud* Olímpio Cruz Neto, "FHC e o fim da Era Vargas," *OLÍMPIO CRUZ NETO* (website), September 1, 2009. https://olicruz.me/2009/09/01/fhc-e-o-fim-da-era-vargas/.

43. Cardoso, *As idéias e seu lugar*.

44. "Neoliberal policies used globalization and its flows to force part of the oligarchic-corporate bloc to execute an escape forward." Negri and Cocco, *Glob(AL)*, 165.

45. Marcos Nobre, *Imobilismo em movimento* (São Paulo: Companhia das Letras, 2013), 65–88.

46. Negri and Cocco, *Glob(AL)*, 129.

47. However, in terms of agrarian reform, the number of families settled during the eight years of Cardoso's government (1995–2002) was 540,000, surpassed by Lula's government (2003–2004), with 614,000, but much higher than the numbers of Dilma's government (2011), with just 107,000 families. Data from INCRA (National Institute of Colonization and Agrarian Reform).

48. Negri and Cocco, *Glob(AL)*, 165–69.

49. In the following decade, Bresser-Pereira converted, without discontinuity, to one of the most "organic" intellectuals of the neo-developmentalism practiced by the PT governments, which succeeded Cardoso after 2003. Bresser-Pereira now leads an influential think thank on public policy, occupied with a core group of economists that develop strategic monetary and political projects.

50. This is another transformation anticipated by Deleuze and Guattari, soon after the Nixon shock, when they claimed that money became "capitalism's true police" (Deleuze and Guattari, *Anti-Oedipus*, 239). "This is money and the market, capitalism's true police." Police, here, in the strong sense of a violent administration of normality, of the everyday ruler that sets the tolerable, the visible and the speakable of social life. In biopolitics, the molecular aspect of power does not mean it is less violent, so much so that the very racism works in flow.

51. For a good synthesis of this transition, see Andrea Fumagalli, "Moneta e potere: Controllo e disciplina sociale," in *La Moneta Nell'impero*, ed. Andrea Fumagalli *et al.* (Roma: Ombre Corte, 2002), 11–40. See also Fumagalli and Mezzadra, eds., *Crisis in the Global Economy*, 151–73.

52. One can say that the acute imbalances of the new governance were not solved during the 1990s, as seen in the new recessionary wave raised by the crises in Asia (1997) and Russia (1998). The overvaluation of the exchange rate in Brazil led to the 1999 debacle, when the abrupt devaluation caused devastation among the companies that operated in the country and reduced the public budget to the point it could not carry out social policies or correct productive imbalances. In Argentina, the collapse of Cavallo's plan and its currency board occurred in 2001, when the last card of Menemism, the confiscation of savings accounts ("corralito"), led to multitudinous turmoil swayed by the cry *que se vayan todos y que no quede ni uno solo!* (Out with them all, and let no one remain!)

53. Mario Vargas Llosa, "El regreso del idiota," *Foreign Policy* 160, May/June 2013.
54. Gonzalez Chávez. "El Dilema Neoliberal De Los Subsidios," *El Día*, January 29, 2012, accessed March 2016, http://www.eldia.com.bo/index.php?cat=162.
55. William, Neuman, "Turnabout in Bolivia as Economy Rises From Instability," *New York Times*, February 16, 2014, accessed March 2016, http://www.nytimes.com/2014/02/17/world/americas/turnabout-in-bolivia-as-economy-rises-from-instability.html.
56. Pablo Stefanoni, "La utopia coreana en los andes," *Rebelión* (author's blog), July 18, 2013, accessed March 2016, http://www.rebelion.org/noticia.php?id=171279. Carlos de la Torre, "El tecnopopulismo de Rafael Correa: ¿Es compatible el carisma con la tecnocracia?" *Latin American Research Review* 48, no. 1 (Spring 2013): 24–43, accessed March 2016, https://muse.jhu.edu/article/504727. Bernardo Gutiérrez, "Ecuador pone rumbo a la economía del bien común," *#CÓDIGOABIERTO_CC* (author's blog), accessed March 2016, http://codigo-abierto.cc/ecuador-pone-rumbo-a-la-economia-del-bien-comun/.
57. For the intricate network of state corruption, drug dealing, and political power in the second biggest Brazilian metropolis, see Luiz Eduardo Soares, *Rio de Janeiro: Extreme City*, trans. Anthony Doyle (Penguin Books, 2016).
58. For comparative purposes, take the number of fatal political victims from the dictatorship in Brazil. From 1964 to 1985, 400 people were killed for political reasons. In contrast, the most recent estimates speak of more than 8,000 indigenous people killed as a result of developmentalist campaigns and land invasions during that same period. See "Relatório Figueiredo," EBC, accessed August 2016, http://www.ebc.com.br/relatorio-figueiredo.
59. On frictions and tensions between developmentalism, sometimes tinted in red, and indigenous ways of life and cosmopolitics, see Salvador Schavelzon *Plurinacionalidad y vivir bien/buen vivir: Dos conceptos leídos desde Bolivia y Ecuador post-constituyentes* (Quito: Abya Yala/CLACSO, 2015).
60. Alejandra Santillana Ortíz, "Cracks in Correísmo?" *Jacobin*, 08.14.2015, https://www.jacobinmag.com/2015/08/correa-ecuador-pink-tide-protests-general-strike/.
61. Before assuming the Vice-Presidency, along with Evo in 2006, Linera was trying to reconcile what, for him, were two reasons which could be reunited in their differences, Marxism and Indianism. *Cf.* Alvaro Linera, "Indianismo y marxismo: El desencuentro de dos razones revolucionarias," *Revista Donataria* 2, March/April 2005.
62. René Zavaleta Mercado, *Lo nacional-popular en Bolivia*, first published by Siglo XXI in 1986 (La Paz: Plural 2008), p. 146.
63. "CONAIE retira bastón de mando al presidente Rafael Correa," *Ecuador Noticias*, April 20, 2011, accessed March 2016, http://www.ecuadornoticias.com/2011/04/conaie-retira-baston-de-mando-al.html.
64. For a thorough work discussing the tensions between governments and movements in Ecuador and Bolivia, which found a resolution in developmentalist terms, see Salvador Schavelzon, *Plurinacionalidad y vivir bien/buen vivir: Dos conceptos leídos desde Bolivia y Ecuador post-constituyentes* (Quito: Abya Yala/CLACSO, 2015). See also Alberto Acosta, *O bem viver: Uma oportunidade para imaginar outros mundos*, trans. Tadeu Breda (São Paulo: Autonomia Literária/Elefante, 2015).
65. Pablo Stefanoni, "La lulización de la izquierda latinoamericana," *Le Monde Diplomatique*, September 2014, accessed March 2016, http://www.eldiplo.org/notas-web/la-lulizacion-de-la-izquierda-latinoamericana.
66. Decio Machado, "Ecuador: la Nueva Disputa por el Poder," *La brecha,* 18.08.2017, http://uninomade.net/tenda/3443/.
67. See Salvador Schavelzon, "El fin del relato progresista en América Latina," *La Razón*, June 21, 2015, accessed August 2016, http://www.la-razon.com/index.php?_url=/suplementos/animal_politico/fin-relato-progresista-America-Latina_0_2292970735.html; Raúl Zibechi, "Hacer balance del progresismo," *Resumen Latinoamericano*, August 4, 2015, accessed August 2016, http://www.resumenlatinoamericano.org/2015/08/04/hacer-balance-del-progresismo/; Gerardo Muñoz, "Notas sobre el agotamiento del ciclo progresista latinoamericano," *Infrapolitical Deconstruction Collective*, October 29, 2015, accessed August 2016, https://infrapolitica.wordpress.com/2015/10/29/notas-sobre-el-agotamiento-del-ciclo-progresista-latinoamericano-gerardo-munoz; and Bruno Cava, "Pueden los gobiernos progresistas sobrevivir al proprio

éxito?," *Lobo Suelto!* (blog), January 6, 2016, accessed August 2016, http://anarquiacoronada.blogspot.com.br/2016/01/pueden-los-gobiernos-progresistas.html.

68. That is what happened, for instance, when Dilma was ousted after massive protests and political scandals regarding electoral campaign funding, which finally led to an impeachment process that decided against the president in both legislative chambers. PT then adopted the *coup d'etat* discourse as the official narrative, trying to build on the collective memory of the Latin-American coups from the 1960s and 1970s. However, by ascribing the fall from power to a Great Other (elite prejudices, imperialism, "economic sabotage," etc.), they fail to address the endogenous reasons for the destituent crisis, which were legion. See Giuseppe Cocco, "La encrucijada de la izquierda brasileña," Nueva Sociedad—Democracia y politica en America Latina, August 2016, accessed August 2016, http://nuso.org/articulo/la-encrucijada-de-la-izquierda-brasilena/.

69. This reading is common within the neo-Gramscian and Laclau-oriented framework. See, for instance, Iñigo Errejón Galván, "La lucha por la hegemonía durante el primer gobierno del MAS en Bolivia (2006–2009): un análisis discursivo" (PhD diss., Universidad Complutense De Madrid, 2012), accessed August 2015. Available at http://eprints.ucm.es/14574/.

70. For the immanent connection between *potency* and *power* within *strategies*, see Toni Negri, *Spinoza et nous* (Paris: Galilée, 2010), 57–81.

71. Schavelzon, "El fin del relato progresista."

72. Giuseppe Cocco and Bruno Cava, "Vogliamo tutto! Le giornate di giugno in Brasile: la costituzione selvaggia della moltitudine del lavoro metropolitano," EuroNomade, October 21, 2013, http://www.euronomade.info/?p=173.

73. See the dossier of articles edited by Giuseppe Cocco, "The Insurgent Multitude in Brazil," *The South Atlantic Quaterly* 113, no. 4 (Fall 2014).

74. See, for instance, Bruno Cava and Giuseppe Cocco, *Amanhã vai ser maior: O levante da multidão no ano que não terminou* (São Paulo: AnnaBlume, 2014). See also Jean Tible et al., "Dossiê: A potência dos pobres." *Lugar Comum* 40, December 2013.

75. Luíz Inácio Lula da Silva, *Carta ao povo brasileiro*, May 22, 2012. *Apud* "Leia íntegra da carta de Lula para acalmar o mercado financeiro," *Folha Online*, June 24, 2002, accessed February 2016, http://www1.folha.uol.com.br/folha/brasil/ult96u33908.shtml.

76. PSOL: Partido Socialismo e Liberdade.

77. James Holston, *Insurgent Citizenship: Disjunctions of Democracy and Modernity in Brazil* (Princeton: Princeton University Press, 2008).

78. André Singer, Os *Sentidos do Lulismo: Reforma Gradual e Pacto Conservador.* (São Paulo: Companhia das Letras, 2012).

79. For Singer, this is a hybrid model that blends elements of Louis Bonaparte, in light of the fiery Marxian reading in *The Eighteenth Brumaire*—i.e., the direct connection between a leader and a disorganized class—with aspects of Roosevelt's New Deal (but lacking the union structure and the workers' struggles), a socio-economic reorganization to face the crisis, in this case, the financial crisis arising from the Asian (1997) and Russian (1998) recessions.

80. The win-win thesis, an illusion sometimes found in further-to-the-left analyses, completely omits the exploitation mechanisms involved in the relative surplus value logic. Generally highlighting the reduction of inequality, it circumvents the fact that capital is essentially a social relationship of exploitation; therefore, the gap between the power of profit and wage may increase without an actual wage decrease. For that, it is enough that the gains in productivity arising from the development of social forces are not distributed on equal terms, which will cause an increase in the exploitation rate, without necessarily meaning a drop in real wages, in absolute terms. Hence, the win-win logic needs to be problematized in light of the dual mechanism between absolute and relative surplus value, which will determine the composition of profit and wages, as well as their mutual relationship. There may be simultaneous economic growth and real wage increase while, *at the very same time*, capital profitability grows disproportionately to wages, thus increasing exploitation. Finally, the reduction of inequality between different income groups does not, in itself, imply a fairer partition of productivity gains, since profitability can turn into accumulation regime, convert into cash, or even circumvent the "national" economy.

81. Although Singer's concept of Lulism seems critical at first, it oddly assumes conciliatory contours when he claims that the reformist pact was the "maximum limit of the correlation of forces" while ignoring any emergent process or sphere of subjectivity outside the party. Hence, the position of the PT becomes the very limit of the possible given those specific conditions. For a good discussion on the ideological character of the concept of "Lulism," see Marcos Nobre, "Pemedebismo e lulismo: um debate com André Singer," in *Imobilismo em movimento*, op. cit., 172–89.

82. This is especially true for Caio Prado Jr. As quoted by Singer himself, Prado Jr. says: "one can still feel essentially all or, at least, the major effects of the Brazilian colonial heritage. We still constitute, in a broad and generalized perspective, . . . a heterogeneous and inorganic human cluster without adequate economic structure, in which the productive activities of great significance and expression are not properly engaged with the specific needs of the population." *Apud* Singer, *Os Sentidos do Lulismo,* 17. As one can see, absence is the analysis driver which Singer explores.

83. Singer, *Os Sentidos do Lulismo,* 102.

84. Giorgio Agamben, *Homo Sacer.* Il potere sovrano e la nuda vita, (Torino: Boringhieri, 1995).

85. For the most thorough work of empirical sociology on the new social composition, although limited to the "objectivity" of sociocultural genesis, see Jessé de Souza, *Os batalhadores brasileiros: Nova classe média ou nova classe trabalhadora?*, 2nd ed. (Belo Horizonte: UFMG, 2012 [2010]). Regarding the production of subjectivity of post-Fordist capitalism as the loading of guilt in the debt-creditor relationship, in a mix of Marx and Nietzsche via Deleuze & Guattari, see Maurizio Lazzarato, *Il governo dell'uomo indebitato: Saggio sulla condizione neoliberista* (Roma: Approdi, 2013).

86. As Bruce Lee would say, be formless and shapeless like water.

87. "Up to this moment [2013], Lulism had managed to maintain a double-sided position: on the one hand, there was the 'statist Lulism,' which promotes a modern, efficient and centralized state management against backwardness, old elites, and corruption as a solution to underdevelopment; on the other, we have the 'savage Lulism,' which counters the Brazilian neocolonial state with the radicalization of democracy, a democracy 'from below,' from the minorities and their becomings. With the 'Journeys of June' and their unfoldings, savage Lulism reconstituted itself independently, outside of the government sphere, breaking the ambiguity." *Cf.* Giuseppe Cocco and Bruno Cava, "Queremos tudo: as jornadas de junho e a constituição selvagem da multidão," IHU Online, August 26, 2013, accessed March 2016, http://www.ihu.unisinos.br/noticias/523060-queremos-tudo-as-jornadas-de-junho-e-a-constituicao-selvagem-da-multidao.

88. Bruno Cava, "La misteriosa curva della retta lulista," Commonware (website), December 3, 2014, http://commonware.org/index.php/cloe/521-curva-retta-lulista.

89. Just think of 2012, when the construction workers of both the mega dams in the Amazon region and the "FIFA-standard" stadiums for the upcoming mega events (the 2014 World Cup and the 2016 Olympic Games) went on strike, throwing a sabot into the gears of developmentalism during the Dilma government. Or else, consider the uprising of the multitudes in 2013, which led to a point of no return, when Lulism and beyond-Lulism trends completely detached from each other, eliminating the possibility of an agreed solution to the conflict between biopower and resistance movements.

90. Paulo Arantes, *O novo tempo do mundo—e outros estudos sobre a era da emergência* (São Paulo: Boitempo Editorial, 2014).

91. See Barbara Szaniecki, "Marée Amarildo: amour et art de la multitude," *Multitudes* (Paris: 56), 2014.

92. Paulo Arantes, "'Democracia de baixa intensidade militariza gestão social'—quatro perguntas para Paulo Arantes," Boitempo (blog), June 20, 2014, http://blogdoims.com.br/legado-da-copa-e-mecanismos-de-repressao-quatro-perguntas-para-paulo-arantes/.

93. Rede Universidade Nômade, "O PT se reduziu a um partido da ordem e pela ordem?," *Lugar Comum* 40, December 2013, 9–15.

94. See the previous section.

95. Up until 2016, the pre-salt layer has still not brought any significant revenue to the government. With the abrupt fall in oil prices, the experts are now in doubt about the economic viability of extracting oil in the depths of this layer.

96. NOBRE, *Imobilismo*, 118–36.

97. Raúl Zibecchi, *Brasil potencia. Entre la integración y un nuovo imperialismo* (Malaga: Zambra - Baladre, 2012).

98. Brian Blackstone and Marcus Walker, "Draghi driblou alemães na defesa do Euro," *Valor Econômico*, October 9, 2012, Wall Street Journal Américas sec.

99. Rafael Correa, "A Europa comete os mesmos erros que nós" *Le Monde Diplomatique Brasil*, December 2013.

100. Lucianne Carneiro, "Os europeus ainda não olharam para exemplos como o do Brasil," *O Globo*, October 25, 2012.

101. Paul Krugman, "Lecciones argentinas para Europa," *El País*, May 16, 2012.

102. Martin Wolf, "Os *holdouts* denigrem os abutres," *Valor Econômico*, September 3, 2014.

103. Soon, though, everybody realized that the dream of a golden future was in fact just the prelude for a long nightmare.

104. Marcelo Neri, *Nova Classe Média*.

105. Giuseppe Cocco, ed., "The Insurgent Multitude in Brazil," *The South Atlantic Quaterly* 113, no. 4 (Fall 2014).

Chapter Three

The Chinese Decade

The Chinese economy came to play a double role. On the one hand, it appears to represent the great and irreversible "novelty" of neoliberal globalization, particularly when our point of perspective is South America. While almost nothing remains of the legacy of the left-leaning regimes, the last South American decade appears to have genuinely been a "Chinese decade." On the other hand, the Chinese advance is seen, especially by voices of the critical globalization studies, as a new "Outside" of Empire, as something that stands for an alternative path. The failure of neo-developmentalism experience in the second chapter takes us directly to the role of China—a true novelty of globalization and of the last decade of South America, and a "Middle Empire" seen as standing "outside Empire." Between these two perspectives—the collapse of the "laboratory" that the left-leaning regimes have constructed, and China as "new-old outside"—the questions that emerge are those of the "moral confusion" that the infinite debt causes and of the increasingly important part played by finance. We will discuss these issues in the fifth chapter, dedicated to currency—to currency and not solely to debt, to currency as open and conflicted reality, to currency as a living thing.

THE CHINESE INFLECTION POINT

About ten years after the great crisis of global capitalism was triggered on Wall Street (2007–2008), debates and conflicts regarding directions and challenges of a new governance model of globalization are still open. Those uncertainties are now cut across by the persistence and unfolding of the crisis itself, particularly in Europe, especially with Brexit, and the Mediterranean basin, and also by the turmoil China is experiencing. After the subprime imbroglio, the crisis moved to the European sovereign debt and now involves, at the same time, in different terms and degrees, the major commodity producers and consumers: China and—as we saw above in chapter 2, Latin America, especially Brazil.[1]

The period of liberal governance of globalization, which was based on the paradoxical assertion of US hegemony together with its discourse on democracies and the free market, is now over. Behind the walls torn down by the exodus of multitudes leaving real socialism, some eschatologists of capitalism believed we could finally achieve the end of history's radiant horizon. The American-style liberalism was replacing socialism as the new and finished teleology of history, in the Hegelian spirit of dialectical reconciliation. The United States appeared as the last and only global superpower, and neoliberalism as universal consensus, the *Washington Consensus*. But the American supremacy would also wear out while new modes of resistance emerged, which implied the development of a new configuration for struggles and the production of subjectivity that animates them. The walls were really down, but it soon became clear that history is far from over. New sorts of walls developed: labor was stratified in new ways and capital achieved new dynamic and flexible frontiers.[2]

The vitality and indetermination of history have evolved along at least two kinds of lines: the modulation lines and the lines of flight. A new *nomos* of the earth, in the Schmittian sense,[3] was articulated around the *modulation* lines of a supranational governance composed of what Hardt and Negri[4] called imperial aristocracy. Regularly gathered in forums such as Davos, this aristocracy of financial global economy operates from a network of supranational institutions for coordination and formulation purposes: G8, IMF, WTO, UN, multinational corporations, banks, as well as risk rating and financial advisory agencies. At the same time, new democratic struggles were creating lines of flight instead of opposing globalization in the name of national sovereignty—in the tradition of the anti-colonial struggles of the second post-war period—or on behalf of a presupposed Outside of capitalism—sometimes occupied by the real socialism bloc, at others by the third-world coalition of non-aligned countries, like the one gathered in the Javanese city of Bandung in 1955. By the end of the last century, those new struggles started building a critical stance within the new material conditions post–Cold War, activating counter-trends and proliferating counter-powers.[5]

That is how the alter-globalization movement asserted itself as an answer to the governance of globalization. Rather than opposing globalization with the entrenchment of national sovereignties according to a statist or nationalist conception, incongruous with the new connective tissue of the world productive fabric, they fought within this very tissue striving to create a new globalized world from the entrails of the existing one. As a result, the World Trade Organization and G8 meetings were now followed by a parallel theater of democratic mobilizations, whose most spectacular moments in terms of volume of direct actions occurred in Seattle, in 1999, and Genoa, in 2001.[6] For the first time since the end of the Cold War, the cycle of alter-globalization struggles unfolded on the "three" sides of the former Iron Curtain: in the West embodied by the United States and Western Europe, in the East—in Prague, for instance, against the IMF in 2000—and in the South, in social world forums organized as a counterpoint "from below" to the World Economic Forum—first in Porto Alegre, in 2000. And that was accomplished without being immediately crushed by Soviet tanks at the hands of dictatorships, in the service of the Warsaw Pact, or tortured by the Latin-American military trained at the CIA school in Panama.

This embryonic new *nomos* of the earth was first intersected by the cutting lines of the September 11 attacks and, then, by the implosion of the lines of flow in the financial crisis of the 2000s. Now, it is again cut across by the lines of the Islamic State (ISIS), in Syria, France, Iraq, or Turkey. Consequently, since the 1990s heyday, the governance of globalization has lost at least two of its pillars. The first was constituted by the expectation of global peace under the American bastion which should follow the end of the Cold War and the financial regulation of the economy. The other was the growth regime of the global economy, which had the inclusion of China as a global industrial driving force at its core. On September 11, 2001, a new kind of terrorism threw cold water on those who diagnosed a new century of *Pax Americana*. Al Qaeda restored the warlords and global instability to center stage. After the attacks on the World Trade Center and the Pentagon, the United States decided to double their bet on military interventionism in the Middle East and Afghanistan region, launching a daring geopolitical security plan under the War on Terror discourse: ubiquitous, pervasive, and network-structured. The strategic reconfiguration of US military doctrine under the Bush administration led the war on terror to a virtually postmodern or "Deleuzian" operation, developing undecidability thresholds and risk gradients in which the military vectors could act from any point at any point.[7] The "lightning war" tactics and the so-called "surgical strikes" of American intelligence and Special Forces in Afghanistan and Iraq were just the first step of this new "smooth" modality of war, which would reach its culmination point with the massive introduction of killer drones the following decade.[8]

In this sense, the rejection of supranational institutions, with the resulting second invasion of Iraq in 2003, led Hardt and Negri to call the US intervention a "monarchical coup," a *coup d'état* by the US neoconservative elite against the Empire.[9] According to Alain Joxe and Giovanni Arrighi, the US debacle in configuring a new *Pax Americana*—which is more than clear in the chronic instability in Afghanistan as well as the growing and potentially catastrophic volatility in the Middle East—alongside the inability to impose their own interest as a global interest, shows that the accumulation capitalist cycle, in which the United States played a hegemonic role, is now in decline[10] and, as we said above, Trump's politics are exactly the opposite of his slogan: he doesn't make "America Great Again," but smaller and smaller. Today, among Europe, the Mediterranean, and the Middle East, the crisis unfolds along uprising lines, wars, and also an exodus of biblical dimensions.

At the same time, once again, the efforts and disputes concerning the construction of a new governance of globalization are doubly intersected by China's presence: first, because China is the globalization major "new fact" and its diplomatic strength has been magnified by its overwhelming economic weight; second, because the country is now going through an inflection which necessarily has and will keep having impacts on global dynamics and on the eventual outcome of the quest for a new world order. "The world's economic centre of gravity has shifted away from OECD economies towards emerging economies during the past two decades." This is the phenomenon called "shifting wealth."[11] Are we back to that China which the Chinese saw as the "Middle Kingdom," as Li Xing describes it, "in terms of both vertical—between heaven and earth—and horizontal perspectives—between the civilized and the barbarians"?[12] That is to say that, at some point, surrounded by enemies that infiltrated the capitalist interstate system, the Chinese civilization *would have decided* to assimilate the instruments of power accumulated by the West, promoting the *aggiornamento* of their ancient and industrious institutions to the highest level of the geopolitical chess game.

Li Xing calls attention to the ironical situation: "China's success lies in the fact that its integration into the world economy has been identified to be a central savior of the capitalist system." However, this success is not exactly associated with the submission of China to the world market, but with a Sinicization of "liberal market capitalism into something new—i.e., 'Chinese socialist market economy,' 'Chinese model' or 'Beijing Consensus'—[when the rise of] the country is immediately demonized as a potential challenger attempting to subvert the capitalist world order."[13]

China is not just the most dynamic and most important emergent economy, but also a diplomatic power of global dimensions: "China foreign policy is already reflecting the economic power attained and its influence is expanding."[14] China is, therefore, undergoing a series of changes and uncertainties that have immediate impact on the global economy, particularly devastating for emerging economies specialized in commodity exports.[15]

Already in 2003, David Harvey was speaking of how dramatic were the prospects of Chinese long-term infrastructural investments and their huge mega-projects. He said that this effort

> is far larger *in toto* than that which the United States undertook during the 1950s and 1960s . . . and has the potential to absorb surpluses of capital for several years to come. It is, however, deficit-financed . . . and that entails high

> risks since if the investments do not return their value to the accumulation process in due course, then a fiscal crisis of the state will quickly engulf China with serious consequences for economic development and social stability.[16]

We know that China's economy survived the violent shock of the 2007–2008 financial crisis, but the questions about return regarding the huge values invested in mega-projects during the anti-cyclical economic politics are still unanswered. When we go from China to the Latin American version of this developmentalist matrix, particularly the Brazilian one, we can see how the choice to accelerate growth through a mega-projects cycle had a ruinous effect, both for the environment and the economy *tout court*.

Michel Aglietta, an economist from the French Regulationist School, and Guo Bai[17] believe there will be no crisis in China, but only growth in a more moderate pace than usual, meaning a slowdown to a more sustainable rate of 3% to 4% per year. Those analysts are betting on an inflection that would enable the country to finally face the inequality issue, going to "phase 2" of development focusing on a more horizontal and productive socialization, improving life indices, and investing in innovation and high-tech industries directly related to their national defense strategy. They come to the same conclusions as the Brazilian developmentalist economist Fiori, who sees the inflection point as a conscious change of course, guided by the strategical thinking of the Chinese party-state, instead of a "forced landing" prompted by the economic cycle exhaustion or related to the "middle-income trap" hypothesis. The slower pace of GDP growth would be compensated by a significant increase in the Chinese household income. China would then be quietly going toward a "new normal," straightening its course for later competing with the United States for leadership in the geopolitical chess game. More than that, some believe that China represents an example to follow or, at least, an alternative civilizational matrix in relation to the American-style democratic liberalism, which won the Cold War. In those lines, Aglietta and Bai wrote, back in 2012, that the broader goal of China is to transform its growth regime. This could then create the grounds for a "Chinese way" out of the global economy current impasse, in the sense that it would still prioritize urban development and tackle environmental issues at the same time. They even believe China may have the conditions to build a "unique sort of capitalism,"[18] something like a new kind of capitalism.

In a more prosaic way, the Chinese government is pointing to the need to focus on activities more consistent with the cognitive capitalism accumulation regime—that is, the production of intangible value.[19] The point is to "move from 'made in China' to 'created by China,'" while transferring "some manufacturing capacity abroad, because here (in China) the labor price is becoming more and more expensive."[20] Indeed, "what we call 'made in China' is really assembled in China, but its commercial value originates from numerous countries and patents which precedes the assembly. It no longer makes sense to think of commerce in terms of 'them' and 'us'"[21] At the same time, nothing indicates we are dealing with a "natural" inflection. This rather seems connected to the contradiction and conflict grounds which are increasingly open: "for many years, household income grew below the GDP, and its main beneficiary was the elite. Changing this involves a political problem, not an economic one, and that is why the adjustments are proving to be so difficult."[22] As we can see, there is no lack of analyses providing for the "possibility of a genuine (Chinese) crisis, which would also mean a world crisis."[23]

For others, China would be in a "1929 moment."[24] The new wave of labor conflicts confirms this other side of macroeconomic turbulence. According to the *China Labour Bulletin* (from Hong Kong), this is apparently a "bigger and more violent (wave) than previous ones." Strikes and protests doubled in the first eleven months of 2015, reaching 2,774 according to the regime's official figures,[25] called by them "incidents." Some say the social pact, "under which the migrants accepted strenuous work shifts and Spartan life conditions away from home in exchange for the prospect of a better future," is wearing thin.[26] Over the past 30 years, the large and increasing inequality was just accepted because the economy was growing at high rates, bringing benefits for the poor. Can what was once "acceptable" be sustained when the economy slows down and people start asking questions?[27] It is not for us here to make predictions, but to point out the extent of the Chinese inflection and the various elements which indicate its great uncertainty and global implications.

Amidst all that, as we said before, sectors of the neo-sovereigntist Left in Europe[28] and Latin America show an odd common ground with right-wing extremists, like Le Pen, in France; Trump, in the US; Salvini, in Italy; or Putin, in Russia. Those nationalistic leftist thinkers seem to be awaiting confirmation of a new *Outside*, to the point of speaking prospectively of the emergence of a "new cold war," which would allow renewing the anti-imperialistic framework. Thus, the change in course due to the end of the progressive cycle in Latin America would represent a counterrevolution against the initiatives taken by emerging countries, particularly Brazil, within the BRICS. In this sense, as Fiori explains, the cleavage which reconfigured the world order did not exactly occur with the end of the Soviet Union, in 1991. A realignment of great impact would have proceeded in almost two decades the fall of the Berlin Wall, and the subsequent dismantling of real socialism in Eastern Europe and Russia. The Brazilian theorist connects this displacement to the Sino-American agreement, which resulted from the meeting between Nixon and Mao in 1972, thanks to Henry Kissinger's diplomatic work. From the Chinese point of view, the rapprochement with the United States occurred in a context of growing sense of siege by the Soviet Union, identified then as their main threat. This was the milestone for the "Four Modernizations" program devised by Prime Minister Chou En-Lai, in 1975, an all-new framing for Chinese development put into practice by Deng Xiaoping after 1978.[29] It is also worth mentioning, from the US perspective, that the impending defeat in Vietnam—mostly by the anti-colonial strength showed by the Vietnamese, but also in part accelerated by the mobilizations on American soil against the war—was one of the main reasons that led the United States to advance negotiations with the Chinese regime.

We can than point to a reconfiguration which aims to contain, from the Chinese angle, the turbulent period of the Cultural Revolution and the Shanghai Commune, as well as the anti-colonial struggles in Indochina, from the American standpoint—both sides allied to permanently surround the common enemy. The turn toward neoliberalism in the 1970s and 1980s—which would break the backbone of the welfare institutions, the European social-democracies, and the left- or right-wing national developmentalism in the South—cannot be dissociated from this change in governance of globalization caused by the new interdependence between China and the United States. Without denying there are tensions and conflicts of interests between those two countries, it seems to us that the lines which characterize the developing *nomos* of the earth are far from matching a reconfigured bipolarity, as if there were a "new cold war" or a renewed clash of imperialisms (or "civilizations"). On the contrary, the tensions in question confirm the polycentric dimensions of capitalism and democracy today, as well as the challenges and dilemmas which the governance of this dense metabolic interdependence faces, quite different from the relative political and economic autonomy of the entrenched dispute between the United States and Soviet Union in the second post-war period.

In his introduction to a selection of Mao's writings, Slavoj Žižek criticizes the lasting grammar of anti-imperialism:

> when one talks about "globalization and its agents," the enemy is externalized (usually in the form of vulgar anti-Americanism). From this perspective, where the main task today is to fight the "American empire," any ally is good if it is anti-American, and so unbridled Chinese "communist" capitalism, violent Islamic anti-modernists, as well as the obscene Lukashenko regime in Belarus (see Chavez's visit to Belarus in 2006), may appear as progressive anti-globalist comrades-in-arms.[30]

Žižek emphasizes that anti-imperialism is a kind of opportunism: "instead of the critique of capitalism as such, of confronting its basic mechanism, we get the critique of the imperialist 'excess,' with the (silent) notion of mobilizing capitalist mechanisms within another, more 'progressive,' frame."[31] We can agree with Žižek: we have previously discussed that the "more progressive" capitalism intends to be, the more statist, nationalist, and authoritarian it becomes, replacing the actual work of struggles' organization and production of subjectivity with the sinister transcendence shortcut. At the same time, the conflict between Trump and social movements, important sectors of cognitive capitalism and big groups of mass media (like CNN) show that there is a real crisis in the relationship between capitalism and liberal democracy, as Žižek and Badiou argue. But also it shows that there is neither shared economy, nor platform capitalism, nor smart cities without real democracy.

CHINA: FROM THE "MIDDLE EMPIRE" TO *OUTSIDE* THE EMPIRE?

In 2000, Antonio Negri and Michael Hardt released a book that, for a moment guided the debate on globalization. *Empire*[32] is now relatively forgotten. The cycle of alter-globalization struggles served as nutrient for that work, which was even considered the first manifesto of struggles for the 21st century.[33] We were living the beginnings of street mobilization also articulated through the Internet, which had a global reach through sites, e-mail lists and independent blogs, like those maintained by the many Independent Media Centers (IMC) around the globe. But this meant more than the use of Internet new tools. This was about a new-shared sensibility breaking free from older certified authorities. The movement motto was "don't hate the media, become the media." In *Empire*, the authors caught up with that Zeitgeist and built an analysis of new forms of supranational, post-sovereign power—that is, "imperial" power—in interconnection with the criticism concerning the sovereignty of corporations.

But the book was not well received within the Marxist Left, above all by those of the third-worldist and anti-imperialist schools, in all their shapes and forms. Even in the area of postcolonial studies (Walter Mignolo[34]) or researches on coloniality of power (Santiago Castro-Gomez,[35] Anibal Quijano), Negri's approach was widely criticized.[36] Arrighi, in turn, observed: "Hardt and Negri presented Empire as a logic and structure of world rule that was in key respects antithetical to the imperialism that Marxists had theorized in the twentieth century."[37] Before that, the geographer David Harvey had already gone in that same direction: "The publication in 2000 of *Empire*, . . . as well as the controversy which surrounded it, challenged traditional debates and suggested that the left-wing opposition had to rethink itself according to a decentered configuration of the empire which showed new qualities, of post-modern nature."[38]

The biopolitical analysis of contemporary capitalism as well as the definition—borrowed from Marc Augé[39]—of empire as "a no-place with no Outside" threw out the window both the criticism of neoliberalism as mere ideology and the anti-imperialist stance. Negri and Hardt's *Empire*[40] set a head-on course against resistance strategies and discourses aimed at mystifying a new Outside in all its versions: *1)* the idea, present among some nationalist or sovereign developmentalist theorists, of a miraculous space which could be free or totally excluded from capitalism, *2)* the reactive refusal of globalization as a mere formal expansion of capitalism in its globally financialized configuration, or *3)* the simple glorification of the third world state to the detriment of the neoliberal western market—as if the market could be organized independently from state power, a conception oddly borrowed from neoclassical political economy.[41]

However, regarding globalization, the most consistent theoretical alternative framework to *Empire* and its ramifications was developed alongside two other lines of thought—first, the world-economy approach to economic history based on the long run (*longue durée*), but also the neo-Marxist analyses regarding the "new imperialism," like that of David Harvey.[42] The latter restate the Leninist thesis of imperialism as the superior stage of capitalism in financialized global conditions. More than criticizing Hardt and Negri, Arrighi and Harvey practically ignored *Empire*, considering it a superficial phenomenon related to the wave of ideological analyses of globalization, which were unable to deal with the class antagonisms within the reconfiguration of global power in the post-Cold War era. Arrighi wrote: "with the collapse of the Soviet bloc in 1989, the buzzword was 'globalization,' not empire or imperialism" and

> the publication of *Empire* in 2000 (did not) significantly alter the situation, for Hardt's and Negri's work simply repackaged and gave a radical twist to the central tenets of globalization-speak, including the proposition that, under the present conditions of global economic and informational integration, no nation state, not even the US, can form the center of an imperialist project.[43]

In turn, the analytical core of Harvey's new imperialism is the molecular mechanism of "accumulation by dispossession"—that is, a renewal of the Marxist concept of primitive accumulation, something that put the Outside at the center of neoliberal capitalism.

According to Aglietta and Bai's formulation, "capitalism is not the equivalent of market economy." The market economy is organized from the social division of labor, socially mediated by the common trust (*fiducia*) built around currency, which makes trade possible. Capitalism introduces a second split due to asymmetric access to currency, which ends up creating a power relationship between capitalists and workers. The difference is in the form of contract: in commercial agreements, the individuals appear as autonomous beings capable of bearing obligations by their own free will; in employment contracts, however, they are forced to accept an unequal pact due to their subordinate and dispossessed situation. Currency is not only a means for the socialization process, but also—and foremost—an end: the accumulation of currency for currency itself.[44] From this perspective, it would be thus a mistake to regard contemporary China as "capitalist," since the workings of currency and labor would not correspond to those traits, which would "only" apply to the Western world. One way or another, the Outside reappears.[45] Then, Aglietta and Bai go on to ascribe the success of Chinese reforms from 1978 on to a fundamental principle: "the *continuity of imperial sovereignty* over two millennia. China is a unitary state central to the perpetuity of the empire." The Maoist revolution and the still current regime of the Communist Party have just "been restoring the legitimacy that was once embodied by the emperor."[46] Thus, the real contribution of Maoism would have been returning to the Chinese that self-esteem once granted by the imperial glory of their ancient civilization, restoring the national sovereignty principal undermined since the West invasion in the 19th century. The Brazilian Left of statist and nationalist tradition is unanimous and enthusiastic with this idea they consider an unshakeable truth: "the Communist Party of China has just updated an ancient tradition by creating some sort of 'Mandarin dynasty,' which keeps on governing China according to the same *Confucian moral precepts* from imperial times."[47] It looks like those leftist analysts actually understand the revolution as nothing more than a restoration and—nobody seems to know why—something "superior" to the Western inter-state capitalist system. And what could that be? The ultimate imperial power, rooted in the bowels of the centuries. This restoration is attuned with the challenges that China's integration into the modern world poses to the party-state regime. That is, the revolution was positive, but only because—as we have said—it was linked to a restoration.

It is not our goal here to rebuild the whole debate, just indicate some of its terms and implications. On the one hand, we have the critical work of political economy trying a different socioeconomic matrix; on the other, there are those which favor culturally biased and exoticizing approaches to China as a "radical otherness." Obviously, there is a vast field of interactions and feedbacks between those two lines of thought. Next, we will focus on two influential theorists: first, we will approach Giovanni Arrighi's stance on political economy; then we will discuss the more anthropological perspective of sinologist François Jullien and his critic, Jean François Billeter.

INDUSTRIAL REVOLUTION *VERSUS* INDUSTRIOUS REVOLUTION

Giovanni Arrighi, in the subtitle of the Spanish edition book about China, assigns to Beijing the "foundations and origins" of the first century of the third millennium and the current situation of globalization.[48] He finds in

Adam Smith and his explanation of market´s logic in *The Wealth of Nations*, the impromptu theoretical description for understanding Chinese development since Deng Xiaoping's reforms in 1978.

Arrighi goes back to the theoretical displacements carried out by Mario Tronti in his youth, in the mid-1960s. The founder of Italian *operaismo* (workerism), breaking with the theoretical and political tradition of Marxism-Leninism, shifted the center of analysis. He first placed Lenin in England[49] and then moved Marx to Detroit,[50] deepening the rupture. By taking Lenin to England, young Tronti would put the whole debate on class, party, class consciousness, and hegemony aside—whether in the terms of Luckács or Gramsci—in order to reaffirm the Marxian method of seeking the contradictions within the highest point of capital development. Instead of looking for the weakest link in the historical and political interconnections of capitalism, according to Lenin's teachings, Tronti prefers to seek for the point where the labor organization is stronger. This inversion is in line with the Copernican revolution which he introduces in the Marxian theory: one must look at class composition to find both the creativity and strength to antagonize the capitalist process, hence exploring first England's factories and then the highly industrialized areas of American Fordism, the quintessential grounds for exercising counter-power within the social relation of capital. By reaffirming the anti-evolutionist Marxian formula in the Introduction (*Einleitung*) to the *Grundrisse*, according to which the man's anatomy explains the monkey's and not the other way around, he broke with every kind of teleology and managed to anticipate the antagonistic subjectivity and, consequently, the 1968 cycle of struggles. "A new era of class struggles is about to rise," he would say in his *Lenin in Inghilterra* (Lenin in England). "The power balance seems stable; the relation of forces is unfavorable. However, where capital domination is more powerful and deeper, the labor threat lurks around."[51]

The *operaisti* were not looking for alternatives on an Outside represented by a Soviet Union about to be de-Stalinized and hopefully returned to the authenticity of the Bolshevik revolution, nor by a China where the revolution would have been renewed by Maoism, nor even by the colonial national liberation wars; they were actually searching for it *inside* the most mature capitalist development. In 1970, though he no longer believed in the political power of labor autonomy and had given in to the siren song of political autonomy, Tronti systematized his theoretical rupture. This time, he put Marx himself in Detroit, not just in the heart of the American Fordist industry, but inside the very driving force of the Keynesian plan and the New Deal arrangements.

By taking Lenin to England, he had solved—refusing them—all diatribes involving "socialism in just one country" and the relations among the different stages of development. He thus sought the renewal of the labor perspective within the most solid link of industrialization. By placing Marx in the United States, he completed this total inversion: it was no longer a matter of understanding the labor struggle within the most advanced point of capitalism, but apprehending the *capitalist initiative* (Keynesianism) in the most advanced point of the labor struggle. "The great initiative of capital occurred only once, and not by chance, after its system major crisis and in the midst of the most advanced labor struggles in its history."[52]

For Arrighi, Tronti's proposal was twice as complicated. On the one hand, by saying the labor struggles were stronger where the "party" (the ideology, the Left) was weaker—in the United States—and weaker where the party was stronger—in Europe—Tronti has shifted all the discussion held until then in terms of the theory of hegemony (Gramsci) and critical theory (the Frankfurt School and its *Situationist* renewal with Guy Debord in terms of the society of the spectacle). On the other hand, while abandoning Stalinism, he reversed the drift by which, "since its founding as a theory of capitalist development and a doctrine of socialist transformation, Marxism had seen its influence migrating relentlessly from the centers to ever more peripheral locations."[53]

As Arrighi reminds us, in "the late 1960s, the epicenters of its diffusion had become the poor Third World countries such as China, Vietnam, Cuba, and Portugal's African colonies."[54] By that time, the *operaisti* (workerists) had already put themselves in a heterodox position in relation to those anti-imperialistic theorists who insisted on identifying an Outside to be preserved, from which a counter-hegemonic perspective would be configured. The *workerist turn* concerning this position was to refuse the Outside, understood as a shortcut, an idealist position—be it the third-worldist version or the transcendental approach to "class consciousness"—in order to place the antagonism in the tension grounds where capitalism establishes its capture and control mechanisms.

In the early 1970s, Arrighi militated in a radical Italian left-wing group called *Gruppo Gramsci*. There, he was part of those who refused to join the post-workerist movement called *Worker Autonomy*, which had Antoni Negri among the leading intellectuals. No one better than him, transferred to an American university after a period of activism in Africa, could explain the discomfort caused by Tronti's proposal, the same one Negrian autonomists would extend in 2000, in the context of the post–Cold War struggles. "Instead of forcing the secret of profit-making, as Marx had promised, the rediscovery (of the labor process centrality) deepened the cleavage between Marxists concerned primarily with Third World emancipation from the legacy of colonial imperialism and Marxists concerned primarily with working-class emancipation."[55] What Arrighi then proposes is to address this gap and, for that, recognize that Marx's work no longer allows solving this paradox. In his own words, "The problem was that *Capital* did provide key insights into class conflict; but Marx's presuppositions concerning the development of capitalism on a world scale did not stand up to empirical scrutiny."[56] That is why Arrighi performs two knight movements of chess. First, instead of Marx, he favors Adam Smith and, second, he puts the latter in a peripheral region (Beijing).

For Arrighi, there would be two kinds of economic development. The first involves the development of productive forces without, however, impacting social relations. In other words, it means the dissemination of market relations conforming to the existing social fabric in a gradual process accomplished by a "multitude of hands." Within this scope, the Marxian law of the falling rate of profit is seen as positive, because it gradually dissolves the monopoly barriers and accommodates social dynamics within optimal distribution of resources. This first type of "development based on the market" represents the *industrious* revolution, grounded on the simple accumulation of capital and the coordination of multiple tasks, with flexibility of means and a reasoning which seeks the best solution for each productive problem. By developing such a concept of "industrious revolution," Kaori Sugihara associates it with the full employment of household labor force and intensive use of techniques based on the experience accumulated since the early days of China, putting to use its civilizational backlogs.[57] The Japanese economist reinforces the argument of strong distinction between the European industrial revolution and another development path—something in which Arrighi finds echoes of Smith's political economy formulations.

Arrighi demystifies some of the criticisms usually brought up against Smith—the fact, for instance, that the "invisible hand" would imply a defense of total self-regulation of the markets, which should be left to their own devices. As a matter of fact, according to Smith, the markets can only function efficiently by means of corrective interventions, especially those which avoid concentration and monopolistic tendencies, in order to preserve and reproduce the material conditions of the market itself. Arrighi then argues that the political economy developed by Smith did not find application in the industrial revolution, which was its own historical environment, but rather in China. Placing Smith not in 18th-century England but in China—with its gigantic market—allows us to approach the industrious revolution with greater relevance.

The second type of development mentioned by Arrighi is the one theorized by Marx and then Schumpeter—that is, the capitalist development which dissolves the given social forms and deeply impacts the lifestyles wherever it goes, leaving a trail of changes behind. This is the kind of development that must expand to escape the overproduction crisis, according to the increased accumulation mechanism, which Marx describes in *Capital*. That certainly falls into the accumulation regime orbit where "all that is solid melts into air," resulting in a continuous revolution of the production means.

Capital runs away from itself, since stagnation would be fatal. Marx places the crisis, instead of balance, at the heart of capitalism, while Schumpeter calls "creative destruction" the process which allows capital to face the crisis—that is, the way capitalism internalizes and puts the crisis to work for its own evolution. This sort of Marxian-Schumpeterean development style proceeds by qualitative leaps, constant clashes, major conflagrations, as well as the ongoing accumulation of capital and positions of power. This would produce the Western singularity, which, according to Fiori, historically generated the capitalist interstate system in Europe. The Industrial Revolution accomplished this "capital-based development," whose promethean dynamism led to the historic consolidation of the capitalist interstate system in the 19th century.

Before becoming a capitalist periphery under the hits of British and French imperialist expansion and its opium wars, China was already a dynamic center in Asia, with a market as big and developed as that of all Europe. This "center," however, was of another kind: first, because it was not linked to the industrial, but to an industrious revolution; second, for its non-imperialist dynamic, for not having the need for continuous growth as in the expanded accumulation regime. Arrighi thus retraces Tronti's formula: where market ideology was strong (the West), the hegemony belonged to capitalism; where market reality was strong (Imperial China), ideology was weak.[58] In this original approach, Arrighi reproduces a tripartite division that he identifies in Fernand Braudel,[59] which proceeds by cycles of accumulation led by a hegemonic power from the military, commercial, and monetary perspectives. The first level consists of the material life and the organization of necessities and demands, the second level by trade relations and mercantile logic, and the third and final one by capitalism and the inter-state system. In this sense, capitalism is one level beyond the markets. Consequently, the market economy can exist independently from the formation of expanded accumulation and the appearance of the modern capitalist states amidst European power disputes.[60]

Due to specific European conditions, Arrighi and Braudel place the emergence of capitalism in that continent, in its complex dynamic of wars and its financial networks, beginning with the territorial disputes of Italian cities in the Renaissance period.[61] It is as if the material expansion and power accumulation would eventually crystalize in a plane of states which compete with each other for hegemony, forming the "modern interstate system." The hegemonic position within this system allows the reordering of the process "from above," on a higher plane where the particular interest of the *hegemon* functions as the general interest of all. When China crossed paths with the expansion of this mostly European interstate system in the 19th century, the country was not ready to deal with the competitive pressures and the Western stage of resource and power accumulation, succumbing in the process. This gave way to a long period under British influence, from the opium wars, in the mid-19th century, until 1945. This does not mean, however, that Chinese economy had no wealth of its own. On the contrary, that is precisely why Arrighi invokes Smith, to explain how the industrious revolution in that empire was not only as productive as the European revolution, but of a different kind. It is curious to note that we can find a kind of re-evaluation of Smith's philosophy also in a libertarian author like David Graeber. Even in this case he makes a distinction between "two" Smiths: one of the *Wealth of Nations* and the other one of *Theory of Moral Sentiments*, where "the pursuit of wealth beyond a point where one had achieved such a comfortable position was pointless, even pathological."[62] Arrighi doesn't make this same shift, but he thinks that Smith is able to grasp the enigma of development much more than Marx's analysis of capitalism.

While Marx attempts to solve the puzzle entering into the *camera obscura* of production, Smith concentrates instead on the analysis of the social division of labor outside the factory. He thus favors the dynamic of productive and knowledge sectors, and what happens inside the factory depends on the size of the markets and not on the internal organization of the production units. The British economist was skeptical about large companies and equity partnerships; he thought rural workers and small entrepreneurs were more important than the process of urban and industrial concentration. Smith's historical sociology would then be more useful to understand globalization, because he focuses on examining the competition between nations within the market, while Marx shifts this competition to the class conflict and the technical transformations into the workplace. What Smith considered a pernicious development path, Marx believed was capitalism, without any consideration at all for the alternatives.

Arrighi dove into the national liberation struggles in Vietnam and in Portuguese-colonized Africa in order to find his own Outside of capitalism, and there he stayed until finding it again, in the first decade of the 21st century, in China. What is even more curious is that he found it in post-Maoist China. Indeed, the mismatch between Smith and Marx leads us once again to the cleavage "Asia *versus* Europe." For Smith, East Asia was not an example of backwardness, but a model for an alternative path. For Marx, however, Asia was the condition which allowed the European way. Based on that, Arrighi believes that Marx wound up omitting the role of the state and inverting the role of primitive accumulation, which would be addressed as the origin instead of the result.[63] According to Arrighi, Europe and Asia had the same dynamic until the capitalist bifurcation, due to the fundamental link among capitalism, industrialism, and militarism.[64] The capitalist militarism—which emerged in the intricate theater of wars and competitive pressures in the genesis of the European interstate system—imposed an advantage of the West over poor countries and countries which developed by means of a *natural path*, creating a spatial stratification of the creative destruction process. This results from the efforts of capital to tilt in its favor the shifts that carry along the "endless accumulation," which marks the unnatural path—that is, the endless expansion of the capitalist regime: "The most persistent and crucial form of primitive accumulation (as a consequence of capitalism's creative destruction) . . . has been the use of military force by Western states to provide the endless accumulation."[65]

According to Arrighi, the last 30 years of growth of the Chinese economy does not mean its assimilation within global capitalism, but the renewal of a cleavage that would separate—if not oppose—market growth and a properly capitalist growth. That is what the Braudelian economist sees as a recurrent sign of the end of an accumulation cycle. The American century would be in decline, as the British Empire once was, giving way this time to Chinese hegemony. The emergence of China then corresponds to the decline of the United States as hegemon, including due to their defeat in the wars after the Cold War, particularly the second Iraq War. It is the quality of human and social capital (health, education, etc.) of cheap labor masses, as well as the diffuse power of the marked economy based on an ancient matrix of productive organization, which explains the formidable level of foreign investment in China. This combination was not forged by foreign capital, but by a process of development based on *indigenous* traditions—including the *revolutionary* tradition that gave birth to the PRC (People's Republic of China).[66] The Chinese diaspora's capital played a decisive role in the intermediation between foreign capital and workers, employees and state-owned enterprises: the relationship between Hong Kong and Guangdong, the alliances between the Communist Party of China and the Chinese companies abroad. Another characteristic of the non-capitalist originality of the Chinese path would be in the fact that the reforms did not include the privatization of state-owned companies. Instead, they opted for exposing those companies to global competition. Foreign capital, Arrighi claims, followed the flows rather than determined them, while export growth only happened as a late episode.

According to Arrighi and Zhang, the economic success of Chinese reforms after 1978, compared to the Soviet Union failure, should be assigned

> not so much to the existence of a large agricultural sector, as Sachs and Woo allege, nor to the gradualism and attention for the common welfare of the reforms, as Stiglitz and others believe. This success should be especially attributed to the fundamental differences between the Chinese and Russian revolutionary traditions.[67]

Unlike the Leninist vanguards, Mao's "mass line" has associated itself with the vast rural proletariat. While forced collectivization in the USSR destroyed the "obshchina"—that is, the communal organization of rural areas, considered archaic if not downright reactionary; in China, those social bases have been valued and became, at least for a while, the driving force of the revolution. Here, we cannot avoid adding that Arrighi is obviously abstracting the catastrophic impact of the Great Leap Forward of 1958–1961, and the millions starving to death because of it. Still according to those authors, the result was the rampant capillarity favored by the modernizing reforms, which spread from bottom up, with local and family cells, with profound and lasting effects. It is worth pointing out that Arrighi's explanation for the reforms' success is diametrically opposed to that offered by Fiori, which, as we have already seen, points to the state centralization (from top down) and the defense of the territorial unit promoted by Chow En-Lai and Deng Xiaoping since 1972, in the context of a strategical and geopolitical alignment with the United States.

Arrighi's research has stamina and presents a general review of the literature on development and underdevelopment, then proposing a new Outside: China today, which—after the two-century-long colonialist and imperialist parenthesis—would have been able to rescue its "originality" via Maoist revolution and post-Maoist reforms. Indeed, as we have seen, Arrighi recovers from Robert Brenner and Smith the distinction between "market economy" and "capitalist economy." The Chinese market economy was not imperialist and it fell into decay because of the colonial and imperialist expansion of the capitalist economy, which broke the virtuous circuits of old China. Therefore, he says, "in spite of the spread of market exchanges in pursuit of profit, . . . the nature of development in China is not necessarily capitalist."[68] What would be missing was precisely the "third term" of the Braudelian scheme, the central instance that overdetermines, from a political and strategic angle, the material life and economic markets. In short, there are two types of development: one corresponds to the *industrious revolution* of Smithian ilk, which is market-based and does not destroy or revolutionize the social world as it advances, and the other relates to the *industrial revolution* of Marxian-Schumpeterian leitmotiv which, in order to make way for capitalist accumulation, cannot proceed unless by "constant revolutionizing of production, uninterrupted disturbance of all social conditions, everlasting uncertainty and agitation . . . all fixed, fast-frozen relations, with their train of ancient and venerable prejudices and opinions, are swept away, all new-formed ones become antiquated before they can ossify, all that is solid melts into air, all that is holy is profaned, and man is at last compelled to face with sober senses his real conditions of life, and his relations with his kind."[69]

By resuming Sugihara's argument, Arrighi states that capitalist development is the reality of an *unnatural* path.[70] Given the threat of collapse under an aggressive external trade, this kind of development is compelled to desecrate traditions and embrace *endless* growth, which is continuous and unwise for it is guided by its own accumulative logic, a producing for the sake of producing, a constitutive delusion. Chinese development, in turn, is based on the market and would thus honor a *natural path* of internal trade and resource distribution. Its growth would have an end, and for that it would be wise. Paradoxically, while industrial development has its own limits—exactly for not accepting any limits whatsoever—the industrious way is more likely to promote new economic developments, for having an end imposed by a strong state, opening the plane of long-term strategy.[71]

Thus, Arrighi concludes, the United States has "created a world market of unprecedented volume and density in which the region endowed with the largest supplies of low-price, high-quality labor has a decisive competitive advantage"[72] that allows it to continue the industrious revolution theorized by Sugihara. This would also explain the fact that the Communist Party of China has come to appreciate the values of Chinese tradition without thereby rejecting the values of the revolution or socialism; the reviewing would have only happened apropos the Cultural Revolution. This could not have been different since the successful modernization coordinated by Deng Xiaoping does not rhyme with revolutionary upheavals of the bases, but fits with reducing the "mass line" to the "just line" of the party-state. However, Arrighi believes the Cultural Revolution also played an important part, consolidating the rural bases of the Chinese Revolution and the peasantry's role, unlike the USSR, where the peasants were wiped out. In this statement, we should again point out that the Great Leap Forward imposed on the peasants a disaster similar to the forced collectivization in Stalinist Soviet Union. Moreover, according to Arrighi, the reforms managed to combine a strategic project with the power of the industrious revolution, which in one way or the other was reconciled with the insertion of China on the horizon of capitalist globalization by the end of the 20th century. The productive factors for that can be identified in the strength of rural/household and municipal economy and in the productive mobilization of migrants toward large manufacturing centers, who would become the country's "new labor class" at the turn of the millennium.[73]

Aglietta and Bai also believe that China represents an Outside, an alternative way to capitalism, and that the Maoist revolution had a fundamental role in building the country's industrialization bases through the mobilization of the rural masses, something no government was able to accomplish until that point.[74] Apparently, the ability to manage the rural exodus promoted by the Communist Party, including the internal passport system, is seen as positive by those writers. The two major ideas of the communist economic policy were, according to Aglietta and Bai, heavy industrialization and a low degree of exchange with the exterior world.[75]

The industrialization history in the Soviet Union and even in Maoist China shows that it was this very theoretical and political bet on the role of the centralized planning state allied with big "national champion" companies, aimed at national geopolitical affirmation, which led those countries to face serious obstacles in the path to development. When it comes to that, Fiori deepens even further his discourse on the singularity of China. According to him, China is different because the state would not be in the service of capitalism. Instead, it is the "capitalist development and the very Chinese state which are in the service of an ancient civilization that is already considered the pinnacle of human history."[76] Ultimately, it is once again the nationalist strategy which improves the position of the pieces on the geopolitical chessboards where, in the end, everything is sorted out: "For the Chinese, the capitalist development is just one instrument more in the defense of its ancient civilization against the successive sieges and invasions of the Barbarians."[77] It is clear that appealing to the "ancient civilization" works here as a transcendence shortcut, a leap of faith that solves the embarrassment of having to explain which would be the "positivity" (or the *end*) of a development in the service of the national state and the status of a party-business elite—a nomenklatura—which capitalized on these "advances" for itself all the way.

However, unlike Fiori and Aglietta, Arrighi concludes his analysis by pointing to the critical phase China was going through in 2005, with an enormous increase in exploitation and the spread of social struggles both in rural and urban areas: "this massive proliferation of social conflicts in rural and urban areas has posed a whole new challenge to the CPC leadership and led to a change of rhetoric and policy towards a more balanced development"[78] between rural and urban areas, between different regions, and between economy and society. When concluding his analysis, to the extent he could develop it, rather than inferring some sort of clash of civilizations, Arrighi predicted an inflection of the accumulation cycle along with the American decline, but this tendency is still open and will depend mainly on how the Chinese government and society will deal with the contradictions and tensions arising from their own development success.

NOTES

1. Amidst this Great Transformation, the multiplication of relationships between China and Latin America is one of the major innovations of this decade. In the 2000s, Chinese commercial flows toward Latin America grew 22-fold.

2. See, for instance, Sandro Mezzadra and Brett Neilson, *Border as Method, or the Multiplication of Labor*, (Durham: Duke University Press, 2013).

3. Carl Schmitt, *Der Nomos der Erde im Völkerrecht des Jus Publicum Europaeum* (Berlim: Dunker & Humblot, 1974).

4. Michael Hardt and Antonio Negri, *Empire* (Cambridge: Harvard University Press, 2000).

5. Michael Hardt, "Porto Alegre: Today's Bandung?" *New Left Review* 14 (March/April 2002), accessed June 2016, https://newleftreview.org/II/14/michael-hardt-porto-alegre-today-s-bandung.

6. Starhawk, *Making It Real: Initiation Instructions: Seattle '99*, PDF, December 1, 1999. Available at http://starhawk.org/Activism/activism%20writings/1999-WTO%20Articles/Making%20It%20Real-Day%202.pdf.

7. Michael Hardt and Antonio Negri, *Multitude: War and Democracy in the Age of Empire* (New York: Penguin, 2004), 41–48.

8. Grégoire Chamayou, *A Theory of the Drone* (New York: New Press, 2015).

9. According to the authors, the coup has completely failed. See the chapter "Brief story of a failed coup," in Hardt and Negri, *Commonwealth* (Cambridge, MA: The Belknap Press of Harvard University Press, 2009), 203–18. "The attempt to create a unipolar order centered on the United States was really a *coup d'état* within the global system, that is, a dramatic subordination of all the 'aristocratic' powers of the emerging imperial order, such as the other dominant nation-states and the supranational institutions, in order to elevate the 'monarchical' power of the United Sates. The *coup d'état* was an effort to transform the emerging form of Empire back into an old imperialism, but this time with only one imperialist power. The primary events and ultimate failure of the coup have by now been thoroughly chronicled by journalists and scholars." Ibid., 205, 206.

10. Alain Joxe, *Les guerres de l'empire global* (Paris: La Découverte), 2012. See also Giovanni Arrighi, *Adam Smith in Beijing: Lineages of the Twenty-first Century* (London–New York: Verso, 2007).

11. OECD/CAF/ECLAC, *Latin American Economic Outlook 2016: Towards a New Partnership with China* (Paris: OECD Publishing, December 2015), 17.

12. Li Xing, "Introduction: Understanding the Hegemony and the Dialectics of the Emerging World Order," in Xing, *BRICS and Beyond*, 1.

13. Li Xing. "Introduction: Understanding the Hegemony and the Dialectics of the Emerging World Order," 14.

14. Alexandre Ratsuo Uehara, "China's Foreign Policy and Bilateral Relations with Brazil," in *China and Brazil: Challenges and Opportunities*, ed. Leia da Costa Ferreira and José Augusto Albuquerque (São Paulo: Anna Blume, 2013), 16.

15. As Camila Moreno puts it, "Chinese dream, developmentalist nightmare." The *Chinese dream*, incorporated within the subjectivity of the new Southern middle classes, is part of a new urbanization, social inclusion, and development matrix whose by-product, inverting the cosmopolitical formula of the Andean natives, is the *Bad Living*, a life burdened with demand for success, psychosocial illnesses, health complications, the ghost of social decline, as well as pollution on drastic levels, unbearable traffic, and systematic violation of rights by the developmentalist state.

16. David Harvey, *The New Imperialism (*Oxford: Oxford University Press: 2003).

17. Miguel Aglietta and Guo Bai, *China's Development: Capitalism and Empire* (Routledge, 2013).

18. Ibid., 200, 399.

19. Yann Moulier Boutang, *Cognitive Capitalism*. See also Yu Zhow, "Making Silicon Valley in China: Zhongguancun," in *The Inside Story of China's High-tech Industry*: *Making Silicon Valley in Beijing*, Yu Zou (Lanham: Rowman & Littlefield Publishers: 2008).

20. Marcos de Moura e Souza, "China não vai se contrapor aos EUA na AL, diz analista," *Valor Econômico*, June 6, 2015. Interview with Zhou Zhiwei, Executive Director of the only Chinese think thank devoted exclusively to studies about Brazil.

21. Pascal Lamy, former Director-General of WTO, on January 22, 2011. *Apud.* Camila Moreno, *O Brasil made in China: Para pensar as reconfigurações do capitalismo contemporâneo* (São Paulo: Fundação Rosa Luxemburgo, 2015), 15.

22. Flavia Lima and Marta Watanabe, "'Preço alto das commodities é como cocaine,' diz Pettis," *Valor Econômico*, September 25, 2015. Interview with Michael Pettis. Actually, Petty says that this is not a problem only in China, but in the United States as well, where "income inequality has the highest historical levels. Moreover, in other countries such as Germany, the share of household income in GDP is very low. . . . Income inequality is approaching the levels of the 1920s."

23. Diego Viana, "Turubulência à vista," *Valor Econômico*, December 22, 2015, The Wall Street Journal Américas sec. Interview with Ha-Joon Chang.

24. Alasdair McLeod, "China's 1929 Moment," Goldmoney/Wealth, November 30, 2001, accessed June 2016, https://wealth.goldmoney.com/research/goldmoney-insights/chinas-1929-moment. See also Frances Coppola, "China's Black Monday Signals the End of Its Growth Cycle," *Forbes*, August 24, 2015, accessed July 2016, http://www.forbes.com/sites/francescoppola/2015/08/24/chinas-black-monday-signals-the-end-of-its-growth-cycle/#69f895f219b2.

25. See "Strikes and protests by China's workers soar to record heights in 2015," China Labour Bulletin, January 7, 2016, http://www.clb.org.hk/en/content/strikes-and-protests-china%E2%80%99s-workers-soar-record-heights-2015.

26. Mark Magnier, "The Wall Street Journal, Shenzhen," *Valor Econômico*, December 22, 2015, The Wall Street Journal Américas sec.

27. Viana, "Turubulência à vista."

28. Boaventura de Souza Santos, "Brasil: A Grande Divisão," *Carta Maior*, November, 5, 2014, accessed June 2016, http://www.cartamaior.com.br/?/Coluna/Brasil-A-Grande-Divisao/32167.

29. José Luís Fiori, "Sobre o desenvolvimento chinês (I)," In *História, estratégia e desenvolvimento: para uma geopolítica do capitalismo* (São Paulo: Boitempo Editorial, 2015), 89–91.

30. Slavoj Žižek, introduction to *Mao: On Practice and Contradiction*, by Mao Tse-Tung (London–New York: Verso, 2007), 5.

31. Ibid.

32. Hardt and Antonio Negri, *Empire*.

33. *No Logo*, from Naomi Klein, is another manifesto addressing the demonstrations of this period which involve the direct actions in Seattle and Genova. See Klein, *No Logo: Taking Aim at the Brand Bullies* (New York: Picador, 1999).

34. Walter D. Mignolo. *Local Histories/Global Designs: Coloniality, Subaltern Knowledges and Border Thinking* (Princeton: Princeton University Press, 2000).

35. Santiago Castro-Gómez. "Ciências sociais, violência espistêmica e o problema da 'invenção do outro,'" and Anibal Quijano, "Colonialidade do Poder, eurocentrismo e América Latina," in Edgardo Lander (ed.). *A colonialidade do saber*. Eurocentrismo e ciências sociais. Perspectivas latino-americanas, (São Paulo: Clacso, 2005).

36. See Gopal Balakrishnan, ed. *Debating Empire* (London: VERSO, 2003). Timothy S. Murphy also discusses the many-sided reception of *Empire* and the second book of Negri and Hardt's trilogy in the last chapter of his *Antonio Negri: Modernity and the Multitude*, (Cambridge: Polity, 2012).

37. Arrighi. *Adam Smith in Beijing*, 175–76.

38. Harvey, *New Imperialism*, 16.

39. Marc Augé, *Non-lieux: Introduction à une anthropologie de la surmodernité* (Paris: Éd. Du Seuil, 1992).

40. In *Glob(AL)*, Negri and Cocco made the same political-theoretical move focusing on the Latin-American perspective, relating it to 21st-century globalization.

41. See Karl Polanyi, *The Great Transformation* (London: Farrar & Rinehart, 1944).

42. Those are not actually criticisms, but research lines that were developed before (Arrighi) and after (Harvey) *Empire*, which ultimately ignored Negri and Hardt's proposal.

43. Arrighi, *Adam Smith in Beijing*, 175.

44. Aglietta and Bai, *China's Development*, 9–10.

45. Fiori also assigns a radical definition of nationalist and anti-imperialistic tones to Chinese specificity (in his case, associated to Russia). "Considering China an 'emergent power' is, at least, an ethnocentric neglect or a serious historical mistake." According to him, it was not the Cultural Revolution, but its definitive closure in the 1970s, in the context of the country's rapprochement with the United States, which created the military and political bases for Chinese state centralization and the strategic program of modernization in multiple fronts, propelling the Asian giant to national development. The rise of China as a world power in the 21st century is the result of this long journey that put an end to the "century of humiliations imposed on China by European barbarism between 1842 and 1945." The arrival of China at the geopolitical chessboard does not mean a simple adherence to the European inter-capitalist system along the Westphalian lines according to which the long duration of Western capitalism was engen-

dered. The Chinese development would actually introduce a new element, represented by its "ancient culture," which would constitute a particular civilizing matrix. The very political-military realignment of the 1970s would be nothing but China's capture of capitalism itself, now subordinated to its ancient civilization as an instrument of defense and projection. See Fiori, "Sobre o desenvolvimento chinês (I)," 91.

46. Aglietta and Bai, *China's Development*, 17.

47. José Luís Fiori, "Sobre o desenvolvimento chinês—II," *Valor Econômico*, March 27, 2013.

48. Giovanni Arrighi, *Adam Smith en Pekín: Orígenes y fundamentos del siglo XXI*, trans. Juanmari Madariaga (Madrid: Akal, 2007).

49. Mario Tronti, "Lenin in Inghilterra" (1966), in *Operai e capitale*, 2nd ed. (Torino: Einaudi, 1970).

50. Tronti "Marx a Detroit," in *Operai e capitale*.

51. Tronti, "Lenin in Inghilterra," 89.

52. Tronti "Marx a Detroit," 290. Tronti quotes the famous remark of economist Rexford G. Tugwell, who participated in the formulation of Roosevelt's New Deal: "on March 4, 1933, the alternative was between an orderly revolution, 'a peaceful and rapid departure from past concepts,' and a violent revolution against capitalist structure."

53. Arrighi, *Adam Smith in Beijing*, 18.

54. Ibid.

55. Ibid., 20.

56. Ibid.

57. Kaoru Sugihara, "The East Asian Path of Development: A Long-Term Perspective," in *The Resurgence of East Asia*, Arrighi, Hamashita, and Selden (New York: Routledge, 2003), 78–117.

58. Arrighi, *Adam Smith in Beijing*, 34.

59. Giovanni Arrighi, *The Long Twentieth Century: Money, Power and the Origins of our Time* (New York: Verso, 1994), 1–27.

60. See Fernand Braudel, *La dynamique du capitalisme historique* (Paris: Arthaud, 1985).

61. According to Wladimir Pomar, Portugal featured one of the historical touchstones regarding the leap from a market economy to state capitalism with the Avis Revolution (1383–1385), which led a dynasty supported by the merchant class to power. After Avis, in the following year, the Kingdom of Portugal concentrated its efforts toward maritime expansion. *Cf.* Wladimir Pomar, *O enigma chinês*, 2nd ed. (São Paulo: Fundação Perseu Abramo, 2015), 11.

62. David Graeber, *Debt*, 399.

63. Arrighi, *Adam Smith in Beijing*, 77 onwards.

64. Ibid., 101.

65. Arrighi, *Adam Smith in Beijing*, 365. "Endless" here means both without end and unwise. We will go back to this issue in chapter 5.

66. Ibid., 351. Emphasis added.

67. Giovanni Arrighi and Lu Zhang, "Dopo il neoliberismo: Il nuovo ruolo del Sud nel mondo," in *Capitalismo e (dis)ordine mondiale*, Giovanni Arrighi (Roma: Manifestolibri, 2010), 201.

68. Arrighi, *Adam Smith in Beijing*, 24.

69. Karl Marx and Friederich Engels, "Manifesto of the Communist Party," Marxist Internet Archive, accessed July 7, 2016, https://www.marxists.org/archive/marx/works/1848/communist-manifesto/ch01.htm#007.

70. In fact, this is a Marxian insight, for whom capital is not only unnatural, but unhuman, strange and even monstrous.

71. Sughihara, "East Asian path of development," 78–117.

72. Arrighi, *Adam Smith in Beijing*, 365.

73. Arrighi and Zhang, "Dopo il neoliberismo," 200–205.

74. Aglietta and Bai, *China's Development*, 60, 82.

75. Ibid., 78.

76. Fiori, "Sobre o desenvolvimento chinês—II."

77. Ibid.
78. Arrighi, *Adam Smith in Beijing*, 392.

Chapter Four

A Chinese New *Nomos* of the World?

THE JULLIEN-BILLETER DEBATE ON CHINA

As China rose to the forefront of geopolitics and the global political economy—not just as part of the workings of the world market and its productive metabolism, but also as a fundamental driving force of its development as a whole—the narratives which insist on a Chinese "radical otherness" began to multiply. In 2006, controversy around the Chinese status regarding thought erupted in the French academic circles dedicated to Sinology, a discipline relatively distant from political discussions. That was the year the sinologist Jean François Billeter published *Contre François Jullien*,[1] starting the controversy against his counterpart featured in the title of his pamphlet. The dissension between Billeter and Jullien had already appeared in some sparse interviews and articles, but Billeter really concentrated his critical fire in the aforementioned book, adopting a provocative posture toward Jullien, whose work already enjoyed great popularity even beyond the field of Chinese studies.[2] This debate concerns us in that it allows the advancement of our inquiry on the Outside as a shortcut for transcendence within the framework of relations and interferences between Latin America and China during the present neo-developmentalist cycle and crisis.

According to François Jullien, while revolution is in the core of Western culture, China is marked by "silent transformations," which replace distinction—proper to difference of kind—with distance. Instead of discreet leaps and sudden ruptures, related to the Western revolutionary tradition, change would arrive from multiple operative gradients, overlapping thresholds of undecidability, and intensive degrees everywhere—not the duality of day and night (Western way), but the endless multi-colored moving nuances of twilight (China). Thus, Chinese thought would represent an alternative and a solution to the Greek transition enigma, which happens to be also a major theme in Marxian tradition: How to go from capitalism to communism from the inside out. This is because such thought converts "the impossibility to mark a transition, its beginning or end. . . . Life and the world, aren't they in *continuous transition*?"[3] The transition is undeterminable from a point of view outside its own process of self-differentiation. "While the logos is a 'definition'—*horismos* as the Greek call it—which draws the limits among characters and attributes in order to identify the Being in them; transition is by definition what prevents us from saying how far a given attribute or quality goes and another one begins." The transition "indeed undoes."[4]

Jullien rediscovers the French post-structuralist authors of the 1960s and 1970s in the core of China's ancient thought, whose approach differs from Western metaphysics. Instead of states and identities, thought proceeds on the basis of transitions, gradients, and thresholds—in terms of *difference in itself*, Deleuze dixit. Jullien illustrates it with the snow, which should be thought while it is melting, from the point of view of its transformation, and this is only possible if we renounce to ground ourselves in the language of being, which would be a habit of Western logos. That is why the "Chinese thought . . . is free to pay attention to the stadium of 'what we see and don't realize': in this stadium where the sensible reabsorbs itself and loses its specificity, . . . where the boundaries unravel. The Chinese thinking would then allow the everlasting transition of things to manifest itself."[5] This continuous "transition" is in fact the absence of transition, the flow which leads us to the "fundamental *harmony* that the Taoists have chosen to call *Tao*: the one whose configuration is without configuration, so *Laozi* says, or whose phenomena is without a particular materiality capable of individualizing it (*wu wu zhi xiang*)."[6]

The Chinese thought Jullien presents is thus a radical alternative to Western philosophy and to those ideas he considers its Greek origins: "going from the *ontological* to the *Taoist* approach it is no longer difficult to recognize, instead of the demand for determination of the *logos*, the validity of the 'vague' and the 'unclear' in the process of things (hu-huang)."[7] Jullien identifies a horizon of harmony and continuity in Chinese thought; in other words, there is no transition that does not correspond to this continuous flow, indistinguishable from life and the world. In this perspective, when we go from Europe to China, the very caesura that invokes the European thinking—from "non-being" to "being"—is reabsorbed, and the genesis itself comes from a *continuation*.

In *Contre François Jullien*, Jean François Billeter challenges the mythology of Chinese radical otherness,[8] which Jullien systematically adopts in his dozens of publications since *Procès ou création* (1989).[9] Julian's objection resumes Edward Said's criticism of orientalism[10] regarding the idea of a disenchanted West whose values are threatened by the enigmatic and disrupting otherness coming from the East.[11] This involves a double operation: the exoticization of the East as a fascinating elsewhere, an object for our aesthetic delight or the foundation of our unreachable utopias, coupled with the counter-exoticization of the West as identical to itself, anchored to whatever notion—say, the "Greek thought" from where the Western values come as opposed to the tradition of the Mandarin scholars from the Chinese Empire. This double symmetrical exoticization also establishes a double generalization as it splits the field of thought between "us" and "them."

Billeter goes beyond the criticism of orientalism in order to ascribe to the other sinologist an ideologizing reading of the authors from the Chinese Imperial era. By distinguishing what would be the ancient Chinese thought from the modern Western one based on the duly translated notions of "immanence," "silent transformation," and "processuality,"[12] among others, Jullien loses sight of the power arrangements from where those discourses derive. According to Billeter, the historical-political grounds for these egregious notions recovered by Jullien is nothing more than the imperial despotic regime, of which the mandarin philosophers partake as employees, with an employee morale. Such concepts present in the neo-Confuncionism system, official ideology of the authorities from the Ming (1368–1644) and Qing (1644–1911) dynasties, would have earned a civilizational pedigree under Jullian's anthropological view. The "Millennial Civilization" and its values, solemnly postulated as a sign of wisdom and even sanctity, are then mere secretions of the very imperial political power. The mandarin ruling class then can latch onto the landscape and claim some sort of dignity to itself while producing a pacified normality.[13] The success of Jullien's theory may also be assigned to the ease with which it engages with an exoticizing common sense regarding China, whose roots are deep in the ground of European modernity, corroboration a perspective that is already ingrained in people—especially the dilettantes interested in Chinese studies.

Billeter suggests that Jullien sometimes joins an old intellectual current that goes back to the Jesuit missionaries of the Counter-Reformation, who exoticized the knowledge acquired during their travels and missions to China in order to safeguard their privilege as gatekeepers of the cultural and material wealth of some far-away civilization, and also includes famous sinophiles like Voltaire or Leibniz. Finally, such self-indulgent attitude Billeter recovers in François Jullien's comparative sinology corresponds to the ethos of the *grandes écoles* from the French Third Republic which, in turn, possess a strange similarity with the old mandarinate.[14]

Therefore, according to the author of *Contre François Jullien*, instead of freeing himself of the epistemic Western-centric bias by means of a crafty antipode of thought, Jullien's anthropology/philosophy of the Outside ends up doing nothing more than to superimpose the voice of the very philosopher by resorting to a self-affirmation of the subject over the matter that should actually drag him beyond himself. The narcissistic self-referentiality of Jullien's work presents the public with a "pretentious ignorance,"[15] a reverential respect for some otherness, which camouflages his alleged stance as privileged interpreter of a remote wealth. In the face of that, Billeter aims at dissipating the fetishism around this mythological otherness and restoring the process of understanding Imperial China to a plane we can actually investigate, and that is the organization of power and its freedoms. Such task becomes even more important in a historical period when China advances within globalization as an inescapable force—something we refer to as a *world made in China*.

Shortly after Billeter published his polemicist pamphlet, François Jullien answered by reaffirming several points of method present in his intellectual work from the previous three decades.[16] According to him, it is not exactly a matter of positing an Outside by exoticizing China, but actually searching for an Outside to displace the understanding of our own culture and thus displacing our own thought. If, as Deleuze and Guattari claim, philosophy occurs in the incessant movement of concept creation, this method helps shake the foundations of our thinking and forces us to develop other concepts when coping with the Outside. Hence, for Jullien, the "other is a tool for a philosophical grammar since at least Plato's *Sophist*," when the otherness is not given in advance, but results from a construction, an effort of thought. Instead of being guided by the construction of an exotic China, one must establish a new point of view, a different perspective for ourselves and our world. Such change in perspective does not imply a change in the *view of things*—since the Western view of things is no different than the Chinese view of things (after all, we all see in the same way, and stating otherwise would mean culturalist relativism)—but a change in the things themselves, as clusters of possible practices. Jullien's perspectivism, which relates to Foucault's post-structuralism,[17] is inseparable from a pragmatism in which the points of view are essentially defined by their possible uses, operations and workings.

Therefore, according to Jullien, the controversy around his work is senseless for it is based on methodological assumptions distinct from the ones he adopts. His work calls into question the concept of cultural otherness and puts in doubt the very possibility of a thought that could stem from itself, as a self-identical. Every tradition is necessarily seen from the outside. Billeter is the one who would actually postulate the existence of an "us" and "them" by assuming there could be a thought attuned to a proper "place of speech," say, a Chinese author speaking of China. Thinking invariably implies exposing oneself to the discomfort of assuming other perspectives, of *moving within one's own place*. When thinking, the self who thinks is another from the very moment it starts thinking. In *The Book of Beginnings*, Jullien states that "we can only dislodge the arbitrary from our way of thought by *leaving* it and, in order to do so, by entering another. But what is this strategic elsewhere that might loosen us from the moorings we cannot envisage?"[18] This "strategic elsewhere," in turn, does not mean becoming the other as if this other could be completely understood, but placing oneself in the threshold that separates the self from the other, that is, in the *differend*. This is an entry that performs an oblique attack[19] on the limits of our own thinking. In short, this is not a matter of "thinking like a Chinese" (something impossible in its own right), but thinking through the friction accomplished by the otherness between China and the West.

Still, since Jullien mentions the "strategic" aspect of postulating the Outside, why not challenge the need to split the operation into poles, into the modern West and the ancient China of the mandarins? Considering the political scenario after the end of the Cold War, where the "clash of civilizations" thesis increasingly spread among several audiences, how can we not suspect yet another thought system based on "strategic" binarism? Transposed to the globalization field nuanced by China (by the *China-effect*), are all Jullien's methodological warnings and refinements sufficient to politically justify this fundamental split between the East and the West? We do believe Billeter makes an important point when he denounces the "return to Confucianism" as the smuggling of the Imperial despotism cult, especially when theorists begin to celebrate the revival of a "clash of civilizations" horizon between American neoliberalism and capitalism with Chinese traits, which the sinologist calls an "inauthentic (historicist) rescue of the imperial era glory."[20] Besides, the "strategic" split which Jullien operates in his work also presupposes that the anti-West posture is good in itself, without a great need for further explanation of actual historical-political processes. Again, by transferring his scheme to the geopolitical or economic analyses, we can identify yet another *a priori*, namely, anti-West, anti-modernity, or anti-imperialist stance as emancipating or transforming perspectives. Thus, as a world-economist like Fiori argues, "in the eyes of the West, the 'Chinese model' is authoritarian. . . . However, the Chinese state has demonstrated an extraordinary ability for self-correction."[21] The Chinese party-state would unmask the democracy as a mere Western ideology in the name of imperialist expansion: "its success reveals there are alternatives to the Western model, which would be a European invention converted into 'historical necessity.'"[22] The French philosopher Alain Badiou, for instance, seems to think the same way: "Today the enemy is not called Empire or Capital. It's called Democracy."[23]

Furthermore, the reading proposed by Jullien finds unexpected echoes, for instance, in many Western analyses looking at the Chinese history for some encouragement for their heterodox approaches to economics: "for a Chinese from the imperial period, human relations (*guan xi*), especially among members of the same family, were at the core of existence." Within this network, "harmony was crucial." Unlike the modern West, intersected by the plurality of power institutions, "in China, the personal virtues of honesty and loyalty, sincerity and kindness, instilled by the family system, *are* the rules which guide social behavior."[24]

It is thus imperial China which would offer us an alternative model comprised of both central authority and centralized management.[25] Central (absolute) authority and meritocracy, as operating principals of bureaucracy, would be two qualities missing from the West. According to Aglietta and Bai, China has solved once and for all the contradiction—which would no longer be anthropological, but only Western—between holism and individualism. The solution is, oddly enough, in the fact that popular sovereignty would be more important than democracy. Still according to those authors, China "has a two thousand year old tradition which instituted the central authority of the emperor. The People's Republic is still based on that tension between the central authority of the Communist Party and society's diversity."[26]

Put simply, the dictatorship of the Qing Dynasty or the Communist Party would be socially pluralistic and for the people, and would work as the "reiteration of absolute central authority, whose legitimacy was based (once and again) on the minimal level of well-being of the population."[27] This is something we can reapply to China and to the impasses of the anti-imperialist and national developmentalist "Lefts."[28] On the one hand, they express their fascination with Chinese development planned by the party-state; on the other, they get restless in the face of American analysts and their disbelief in "democracy as a universal value," as we can see in this revealing comment by Fiori: "the democrats of the world, particularly those on the outskirts of Europe and in Latin America, better watch their backs."[29]

It is interesting how Marxist thinkers and heterodox economists end up converging with philosophical approaches such as Jullien's, which tries to solve the Chinese revolution episode by claiming that "the Chinese thought did not have to—or could not—produce any rupture of modernity as Revolt." After denying to China the "tradition of rupture" proper to modernity, Jullien goes on and says that "the authority of *The I Ching or Book of Changes* was never called into question; and it was from Europe that China had to borrow its contemporary modernity and, above all, the Revolution."[30] However, in another moment, when speaking of the attacks on the WTC and Pentagon on September 11, 2011, he says "Could we seriously imagine that, with the fall of the last Berlin Wall and all nations then united, the negative would permanently withdraw from History on its tiptoes, a History reaching its finale, once and for all pacified?" And then he asserts that no, there is no exclusion of the negativity. On the contrary, for the French sinologist, it remains fully within history: "this negativity that we find throughout history and never disappears could no longer be relegated to the outside—mundialization suppressing this Outside."[31]

So, for Jullien, China is not anymore an outside but the (best) way to be inside because of its traditional thought that refuses to think in terms of discontinuous ruptures and differences of kind. According to the French Sinologist, globalization itself is the ground for "*global* transformations" which can no longer accommodate actions and events with sudden transitions, because "economic interests are now too intertwined in the same 'world' by the laws of the market."[32] Besides, globalization is the result of communication, which could do what religion has only done in part: "Which country, which culture, which lost place, including those in China, still escape (it)?"[33]

In short, with globalized capitalism, there is no Outside anymore. Discussing contemporary China after the Maoist reforms, Jullien says that

> Deng Xiaoping was the silent transformer of China: advancing step by step, brick after brick, more than devising some plan or model, but without recurring to empiricism (pragmatism), which is the opposite of our idealism, for it involves tenaciously attaching to the concrete, tapping into . . . the self-development capability of the processes.[34]

Jullien views wars and revolutions as small drops in the ocean. He claims that "contemporary China managed to turn its social and economic system upside down through a continuous transformation, leaving the regime and the Party in power, but deeply transformed." Unlike what happened with the USSR, which became Russia again, "the end of Maoism was not accomplished by the collapse of Mao's statue, but by a prudently managed silent transformation under the Great Helmsman image (from Revolution leader to founder of a new Empire)."[35] Once again, the ancient empire and the Outside that have just been denied come into light.

The narratives on China's economic success place this empire in a "totally ahistorical conception, in which neither culture nor society change over time."[36] As Billeter points out, "in order to build parallel histories that validate one another, (the *comparatist* intellectuals) neglect all that is discontinuous, contradictory, or incomplete, in short, every problematic aspect of all sides."[37] One of them is clearly related to the "myth of *one* China, always equal to itself and different from the rest of the world."[38] In reality, those successes are intersected by antagonistic tendencies and new conflicts—as the encampments irradiated from the Tiananmen Square protests (1989) and, more recently, the Umbrella Movement (2014) in Hong Kong have shown.[39]

BETWEEN THE NEEDHAM PARADOX AND TIANANMEN SQUARE

The debate over China as an Outside in a way brings us to the crucial debate over Eurocentrism and ethnocentrism. The competition along the East/West *line*, which Carl Schmitt regarded as definite, is no longer that crucial and, at the same time, was not replaced by the North/South cleavage. Similarly, the decline of US hegemony does not necessarily mean the rise of another counter-hegemonic sovereignty located in one other state (China, for instance).The Marxian proposal of entering the *camera obscura* of production is still valid, but we need to enter other production spaces and times and seek the new struggles and horizons of the possible *within* the new production relations. In other words, in the West as in the East, in the South as in the North, in the factory *and* in society, in production *and* in reproduction, between the South becoming in the North and the North becoming in the South what has undergone a transformation is the very *camera obscura*, which is no longer the factory or the office, but the metropolis.

In this sense, it is perhaps a matter of changing the perspective and assuming an anthropological point of view. Jack Goody's analysis of how Europe imposed the narrative of its past on the rest of the world helps us in this regard. The ethnocentrism is not in denying the specificity of the other—in this case, the Far East—but in declaring it a mirror of the Western (European) development uniqueness. By adopting the anthropological perspective, we can regard "the development of human society from the Bronze Age . . . in different terms, as an ongoing elaboration of urban and mercantile culture without any sharp breaks involving categorical distinctions of the kind suggested by the use of the term 'capitalist.'"[40] Still according to Goody, "if capitalism marks all societies, its dimension of Western uniqueness disappears, taking away the problem of its explanation." The only remaining question—which is no little thing—would be "explaining (its) increasing intensity," maybe with the "abandonment of the term 'capitalism' altogether.'"[41]

The real turning point would be the urban revolution, and this was a worldwide phenomenon from the start, not just a European one. With that, Goody is trying to go beyond the concept of "modern rupture," devised and proposed by Joseph Needham. According to Needham, the rupture in the West occurred with Galileo's split, when "science merges with mathematics and becomes universal," while the other innovations until then remained restricted to ethnic areas of birth, in a context of mutual incomprehension. However, he says, the ancient world must be thought of as a totality, which also includes Africa. We then have the paradox framework: "Why—on grounds of this totality—did modern science as the mathematization of the hypotheses concerning nature, with all its implications in the context of advanced technology, make its rapid advances only in the West, at the time of Galileo?" The paradox soon unfolds into another question: "Why, between the 11th century B.C. and the 16th century A.C., had East Asian culture had been able to apply the human knowledge of nature to useful ends in a much more effective way than the European West?"[42]

Goody proposes going beyond the "Needham paradox"—that is, he tries to surpass Needham's explanation of why, despite the initial advances in China, it is the West, and not the East, that opens the way to modern science. A question David Landes also raises: "Why industrial capitalism was born in the West, particularly in North-West Europe, and not in China, which a few centuries before was the richest and the most politically, economically and technically advanced country?"[43] A good way of going beyond the Weberian explanations—in terms of the spirit of capitalism—is to develop them from inside out. Rosenberg and Birdzell do that by saying that, before Needham, it was Weber, by means of the protestant spirit, who tried to explain the "success" of the West in the face of a far more advanced China. However, the Protestant Reformation was more the result than the cause of the capitalist revolution of economy in Europe. Based on Joseph Needham's work, they point out that the Middle Empire had, for a long time, superior techniques in relation to the West and a body of educated employees, recruited based on merit; at the same time, the very fact that they had this bureaucracy so early on, and therefore a highly centralized rationality of power, has resulted in China's setback. In turn, the Western "backwardness," or the fact that they lacked this bureaucracy, allowed—still according to Rosenberg and Birdzell—a decentralization of power and decisions, which was crucial for the systematical employment of inventions and technical mechanisms in economic development.

Even more interesting is the fact that here we have elements of the Western difference which deny its uniqueness at the same time: "From the Chinese example, we are able to formulate the hypothesis that the late development in Europe of something equivalent to the mandarinate—the public service—contributed by leaving an open field to the emergence of capitalism."[44] The European backwardness—that is, the late arrival of a bureaucratic body—allowed such an "advance," and there would not be much of a difference between "mandarins" and "public service" according to Needham. From that perspective, the "enigma which established in China the combination of advanced technology and economic stagnation is just a particular aspect of a more general issue regarding the relations between political structures of the imperial kind and growth."[45] What we have here is an opposite perspective regarding the developmentalist approach. The imperial aspects are now seen as a problem and not an advance; conversely, the power dispersion during the feudal period is considered an advantage over Chinese centralization. It may be possible to think that the current Chinese inflection will come across a series of traps regarding the renewal of this same impasse.

This idea is reinforced when we remember the two great illusions about capitalist development pointed out by Rosenberg and Birdzell. The first of them is believing that the "wealth of the West" would result from the mass production accomplished at the factory, which causes the third world countries recently freed from colonial domination to wish for a fast growing industrialization, as was the case of the USSR in the first half of the 20th century and Maoist China in the second half. In reality, the industrial dynamic in the West has only set in *after* the development of trade, mercantile generalized relations, and its corresponding institutions; besides, the factory has never been the main employer for Western workers. The second illusion is turning the result of the industrialization process into its premise, as if taking a "still out of a movie" without the action which corresponds to it. Capitalism, as they emphasize, is really this action.[46]

We can extend this approach and displace the position between two types of development. This is a necessary move in order to avoid overlooking the whole history of the East[47] and also to prevent keeping the debate trapped by the "progressive" means of measuring advances and setbacks. Goody, for instance, calls the dualism between modern and early science into question when he presents a different perspective to the European dynamic after 1600. Instead of a revolution, he proposes approaching it as an "*acceleration* with which these styles both developed and interconnected"; in other words, as an acceleration of tendencies without the occurrence of an actual large-scale qualitative innovation, except for the probabilistic notion, which was indeed a European invention.[48] Asia and Europe have a common history, that of the urban revolution of the Bronze Age.[49] The great civilizations from both the East and the West have a common heritage, which varies only in terms of different velocities and their variations (accelerations), without configuring a separation between the West and the East or any other dichotomy as well as some kind of exoticizing Orientalisms. As Goody suggests in the title of another book of his, we must grasp "the East *in* the West."[50] The Italian Renaissance, a crucial moment of European transformation, is related to the coming together of Europe and Asia through new modes of communication, particularly those fostered by the shipyard and naval revolution and the press.[51] The revolution is denied (there would only be accelerations), but it soon reapers in another context: the technological one (navigation and press). However, Goody tries to overcome the ethnocentrism by systematically criticizing the idea of Western *uniqueness*. In order to do so, he proposes an approach that also avoids its use from inside out. According to him, "a hidden ethnocentric risk is to be Eurocentric about ethnocentricity, a trap postcolonialism and postmodernism frequently fall into." Then, "if Europe didn't invent love, democracy, freedom, or market capitalism . . . , it did not invent ethnocentricity either."[52]

From this perspective, criticizing ethnocentrism implies the *refusal* of the postulate of an *Outside*. It involves identifying the continuous interferences and overlapping elements of Western and Eastern culture and history, therefore recognizing Western specificity as nothing but a particular and temporary phase, as well as the result of stratifications and successive switching of perspectives. The ability to achieve capitalist development was not the sole prerogative of European economies; they were all capable of it. The West had only one relative "advantage," Goody says, in the same sense as Rosenberg and Birdzell, and it was the *backwardness* in relation to Islam and China. More than that, we should abandon the very notions of "advantage" and "backwardness" and, in order to do so, we must no longer separate science and savage thinking, but consider their historical continuous hybridization.

The "birth of science is not at all contemporary to the European Renaissance for the simple fact that it already existed." We could then say that the distinctions between early and modern science, or technology and science, are not the basis for European cultural "superiority," but its result due to a *domination* effect: "The Needham problem, posed as such, does not exist." The question at hand is of another kind; it refers "to whether European primacy in terms of modern science is to be regarded as an undisputable fact."[53] In other words, it concerns the domination and power relations as well as their productive aspects. Within those relations, China has not constituted an Outside, much less today, not least because the West itself is no exception: "Today, in fact, what is put into question is the scale of the Western 'revolution.' This may have 'happened in a less radical way than was thought.'"[54] Contemporary China is neither sheer restoration of the Middle Empire nor something "outside" the Empire.

THE BLACK HOLE OF THE MAOIST REVOLUTION AND THE OCCUPY TIANANMEN SQUARE

One of the most curious features regarding the analyses of China's role in global governances and economy is the treatment given to the Maoist revolution and two more inner events to the communist regime it established: the Cultural Revolution and the Tiananmen Square protests. Basically speaking, the more an analysis identifies itself with a progressive or left-wing framework, the less likely it is to mention the encampments of Tiananmen Square and to solve the issue of Maoism, diluting it in the long-term historical perspective—that is, the "Kingdom of the Middle" perspective. In general, a few (positive or negative) words about the revolutionary period, its (national) conquests, and its (cultural) excesses conclude the subject. Regarding the Tiananmen Square Massacre—the first movement to present the struggle methods later found in Seattle and Genoa, but also in Tahir Square, in Cairo; in Puerta del Sol with the Spanish 15M; in the Occupy Wall Street movement; in the Ocupas in Rio de Janeiro, São Paulo, and Fortaleza; and in the Nuit Debut movement, in France in 2016—the issue is dealt within a few words of grief for its "tragic" outcome (it is unfortunate what happened, but . . .) on June 4, 1989, with hundreds, perhaps thousands of dead, killed by state violence.

Paradoxically, the liberal texts are the ones that provide more room for the analysis regarding the Maoist role (including the Cultural Revolution) and what really was the Tiananmen revolt. An example is Henry Kissinger's book on China, where we find a whole chapter—besides certain solidarity with the regime—with a long and detailed analysis of the material factors involving the student movement of Tiananmen Square. Kissinger's thorough analysis shows us there were material dynamics at the basis of the Tiananmen commune which are comparable to those found in the Brazilian uprising in June 2013, particularly a relative improvement of access and living conditions of the poor. Another similar aspect is the fact that compared to the Soviet Union, where a collapse was imminent, "China seemed stable."[55] Well, compared to the Venezuelan debacle and the Argentine crisis, Brazil also seemed the most stable country in Latin America until it was not, as it has actually never been.

A good opportunity to take stock of Maoism and contemporary China is presented in Slavoj Žižek's introduction to a series of essays by Mao Tse-tung, which we mentioned above.[56] It is not that Žižek offers a historical reconstruction of events that would allow us to do so, but for the overview he traces based on his thoughts on Mao's texts. Indeed, the introduction is extremely dense and gives us a lot of elements for thought. At first, Žižek places his considerations about Maoism within the general debate on "the moment of the Fall, when things took the wrong turn in the history of Marxism." Then, he clearly refuses this trap, stating that "the Fall is to be inscribed into the very origins." It happened in "two great passages or, rather, violent cuts: from Marx to Lenin (and) from Lenin to Mao." Marx expected the revolution to occur in the most advanced countries under the leadership of the working class and it "took place in . . . (the) wrong countr(ies)": first in underdeveloped Russia—where the industrial workers constituted a little minority—and then in China, the country of great masses of poor peasants.[57] Mao mobilized the "Asiatic radical strangeness" within the Chinese radicalized version of the communist movement, which began with Lenin.[58] However, the introduction is organized around *one* major critical statement from which all other considerations derive. Mao's problem would be in the fact that he did not understand the importance of the "negation of negation" and because of that he was unable to see that the "negation of negation" was the only way for the revolution to be victorious. "The conceptual consequence of (the) 'bad infinity' that pertains to vulgar evolutionism is Mao's consistent rejection of the 'negation of negation' as a universal dialectical law."[59]

"True" victory is when even the "enemy" starts thinking like the victorious forces—that is, when the enemy changes his values and enters the revolutionary spiral. "The true victory (the true 'negation of negation') occurs when the enemy talks your language."[60] Because he did not understand how the negation really works, Mao was unable to turn the revolution (the negative) into an alternative (the negation of negation) and ended up wanting to solve the revolution impasses through continuous renewal, in particular with the Cultural Revolution event: "all temporary stabilizations of the revolution amounted to so many restorations of the old Order, so that the only way to keep the revolution alive was the 'spurious infinity' of endlessly repeated negation which reached its apex in the Great Cultural Revolution."[61] Mao could not see that the negative dynamic is in fact the dynamic of capitalism, which is characterized by its ability to continuously revolutionize itself. According to Žižek, it is no coincidence that, with Mao's successors, the revolution became the last throes of the system it wished to overthrow. "So, in a way, there is a kind of poetic justice in the fact that the final result of Mao's Cultural Revolution is today's unprecedented explosion of capitalist dynamism in China."[62] In others words, there is a connection between revolution and capitalism, and this is especially relevant in Maoist China: "There is thus . . . a profound structural homology between Maoist permanent self-revolutionizing, the permanent struggle against the ossification of state structures, and the inherent dynamic of capitalism."[63] Žižek unfolds this general statement into at least two lines of thought: on the one hand, he first recognizes the tragic aspects of the forced industrialization promoted by Mao and then says that the millions who died of starvation from it are no reason to compare Chinese communism to Nazism; on the other, he develops a more general criticism concerning the alter-globalization movements because they fall into this same trap and fail to prevent the capital from assimilating their potentialities within its dynamic.

Our concern here are those remarks regarding the "power" feeding on counter-power; in other terms, capitalism feeding on the struggles that were supposed to bring it down, as if exploitation and the struggles against it were made of the same homogeneous matter. Žižek adopts three references to do so: Boltanski and Chiappello's book on the "new spirit of capitalism"; the employment of concepts from Deleuze and Guattari by an officer of the Israeli army to describe war tactics in the occupied Palestinian territories; and finally, the transformation of post-Maoist China into the heaven of late capitalism.[64] We have three interconnected issues here: First, this association of the impasses and developments of Chinese communism to some theoretical mistakes of its leader; second, the relationship between capitalism and its spirit; and third, the question of knowing if the capitalism established on

China's terrain is really "the" late capitalism or just one of its many manifestations.

We do not believe that the return of capitalism in China has been the result of Mao's inability to understand the true dialectics of negativity and therefore his inability to grasp the necessary "negation of negation." On the contrary, the biggest problem was his ambiguity in breaking with the Diamat. After asking "what is synthesis," Mao answers: "You have all witnessed how the two opposites, the Kuomintang and the Communist Party, were synthesized on the mainland. The synthesis took place like this: their armies came, and we *devoured* them, we ate them bite by bite."[65] Mao then goes on, showing an aspect of his thought which is utterly negative and authoritarian: "One thing eating another, big fish eating little fish, this is synthesis." According to Žižek, this synthesis "occurs not when the opposites are united, but (only) when one side simply wins over the other."[66] There is still a second dimension to Mao's ideas, one that is potentially affirmative and demonstrates an anthropophagic intuition: instead of a big fish eating little ones, all fishes exchange their points of view. Then, when Mao claims "there is not a single commune, a single *hsien*, a single department of the Central Committee, which one cannot divide into two,"[67] his problem is not, as Žižek believes, that he "fails . . . to proceed to the properly Hegelian 'identity of opposites,'" but that he fails to really fly away from Hegelian dialectics. Only a post-Marxist *politics of multitudes* can accomplish this flight, and that demands the constitution of a terrain for radical democracy. A line of flight that may actualize, in very new terms, Maurice Merleau-Ponty's proposal of a "non-communist left."[68]

From this perspective, Mao's problems and limits are the same as all "communist" revolutions and their respective socialist—state centered—regimes. When in Moscow, Marx became Lenin and, when in Beijing, Lenin became Mao: both were Stalinists. When in underdeveloped and feudal Russia, Marxist criticisms concerning the anatomy of capitalist humanism and bourgeois democracy was converted from an immanent plateau of constituent productive forces into a transcendent horizon of constituted relations of production. In semi-colonized and poor agrarian China, socialist forced industrialization immediately showed its dramas and limits without having to go through a period of war: after a false attempt on democratic renewal (with a Cultural Revolution that did not touch Mao's image of transcendence), Maoist China explicitly assumed the industrial project as its model—as the Soviet Union had done. However, differently from the USSR, it did not do so from the outside of Western Fordist hegemony, but right within its crisis and neoliberal restructuration. In this sense, we can say that real socialism, or Stalinism, is the next stage of socialism and the superior reason of capitalism, the rationalist culmination of Hegelian Enlightenment. Ergo, this is the closest one gets to reactionary revolution endorsement and, at the same time, a cosmopolitics whose "elements"—as described by Emmanuel Levinas—wear us down through society and state, by means of an "end of the world without final judgment."[69] Capitalism does not revolutionize itself, it just "runs after" the revolution and its "march of freedom."[70] This is the dynamic, coming from within and against capitalism, that allowed it to come out as a "winner" in a competition (with real socialism) in which it was not the main competitor.

Regarding the relationship between capitalism and the spirit of the times, Žižek employs the work of Boltanski and Chiappelo without stressing that this comprehensive research on the discursive regime of neoliberal capitalism, despite the advances it promoted, has two major problems. The first one is repeating Max Weber's approach and attributing to the "spirit" (the Protestant Reformation) a material role of linear causality that many historians have challenged over and over, pointing to the fact that material transformations were at least as important as the religious aspects and that the "reformation" was more a result than the premise for a capitalist development whose genesis can be traced to Italian catholic cities. The second problem stems directly from the first: by operating this inversion between material transformation and mentality change, Boltanski and Chiappello wind up turning neoliberalism into an "ideology" that could be defeated as such and—even worse—separating and even opposing the "social" and "cultural" struggles. Thus, the authors completely ignored the critique concerning the cultural dimensions of capitalism developed by Guy Debord and other *Situationnistes* of the sixties and employed all literature on immaterial labor, post-industrial capitalism, and general intellect as if it were part of the new management processes of flexible capitalism, with its lean factories and Japanese methods. At the same time, they attribute the weakening of the struggles to the predominance—after May 1968—of cultural movements (of the alternative culture) over social movements (union-like actions, for instance). Thus, Boltanski and Chiappello failed to grasp the main point, that the "spirit" of contemporary capitalism is new not (just) because it changed in relation to the protestant ethics and the Taylorist large industrial society, but due to the fact that capital now also accumulates on the very immaterial and cognitive landscape (on the grounds of the intangible and the sharing relations of goods and knowledge). They also did not see that the new cycle of struggles and movements would come from this new quality of a work that no longer belongs on the factory floor, but to the metropolis, in the wake of those "cultural" movements which they saw as a diversion that weakened the ability to resist. Contrary to what they claim, the crisis lies within the traditional "social" dimensions of criticism, as well as in their union and representative modes, while social struggles and the fight for democracy find their innovative ground in "cultural" criticism. More than that, the cleavage itself is no longer relevant. This appears when Žižek mentions Brian Massumi as someone who clearly formulated the idea "that contemporary capitalism has already overcome the logic of totalizing normality and adopted the logic of erratic excess." In Massumi's own words: "The more varied and even erratic, the better. . . . It's not a simple liberation. It's capitalism's own form of power. . . . (I)t's capitalism's power to produce variety."[71]

By evoking the Italian Workerist Marxism, the Foucaultian criticism of power, and the Deleuzian concept of "line of flight," we can oppose this approach with the idea that the "production of variety" is not capitalist, but only captured by it. Brian Massumi claims that "capitalism starts intensifying or diversifying affect, but only in order to extract surplus-value. It hijacks affect in order to intensify profit potential." When doing that, he substantializes the very notion of Capitalism and fails to grasp the new production relations and their contradictions. Within this approach, anti-capitalism becomes a moral transcendentalism, between impotence and an authoritarian option. Capital may value affects as "political ecology, ethical field of resistance to identity," but the production of life (subjectivity) and resistance are not capitalist themselves because they came "first."

Following those observations, we arrive at the third issue we would like to discuss. Contemporary capitalism is certainly global, but its devices are rearranged under new terms. To capture diversity it is necessary to produce singularities and to stay within the subjective production of "erratic excess." This new configuration of capitalism and its mechanisms to capture values are different from the industrial-Fordist model, dominant in the West during the second post-war period. Its global sourcing and multitude of outsourcing networks do not cover the world homogeneously, as globalization has engendered a smooth space, but by modulation of heterogeneity and new kinds of international division of labor. On the one hand, China's integration into the new global capitalism is part of the capitalist process of inclusion of all social life, especially the poor; on the other, China's specialization in industrial production and low-wage labor cut it from the more advanced forms of excess accumulation in cognitive capitalism.

There is one other dimension to the Chinese situation, manipulated by the revolutionary rhetoric and/or suffocated by the more traditional system of repression, which we must take into account in order to open the horizon of possibilities. Jean François Billeter, for example, proposes an original approach to the Chinese Cultural Revolution. He views it as a process fostered by Mao Tse-tung due to his marginalization in the face of growing criticism around his leadership (particularly after the Hundred Flowers campaign), which was capable of asserting his autonomy, but also got out of control and had to be contained by state violence (led by Mao's ally Lin Biao). Billeter claims that

> Mao plays with fire as he risks seeing this (student) rebellion converted, here and there, into an authentic movement of social and political emancipation, especially if young workers decide to join it and he has to crush, with blood when needed, the beginning of a true revolution that he will have triggered. That is what he will end up doing in 1968 with the army's help.[72]

Jacques Rancière, who mentions 22 times the Cultural Revolution and violently criticizes it in his *Althusser's Lesson*, from 1974,[73] develops a similar (and partially self-critical) approach in the "Foreword to the English Edition" of 2010:

> We cannot be satisfied, today any more than yesterday, with the inverse thesis, which essentially reduces the mass movements of the Cultural Revolution to a simple manipulation carried out by Mao Tse-tung. To recover a power he had lost in the apparatus of the Party. But it is also equally impossible to justify the zeal with which we tried to validate the official image and discourse of the Cultural Revolution.[74]

Before that, the "Great Leap Forward" (between 1959 and 1961) that Mao imposed on the peasants, forcing them to adopt "small blast furnaces," had caused even more famine than the notorious widespread starvation generated by the socialization of rural property in Soviet Ukraine between 1930 and 1934. Thanks to Jasper Becker's work we now know that 35 to 45 million Chinese died during that period.[75] According to Žižek, "If one is to believe Mao's latest biography, he caused the greatest famine in history by exporting food to Russia to buy nuclear and arms industries: 38 million people were starved and slave-driven to death in 1958–1961."[76]

Despite the rhetoric, we know that socialist countries manipulate and hide information; those data are then prone to be dismissed as anti-communist products. That is what many said in the West against the Soviet dissidents who tried to raise awareness within the international communist movement regarding the massacres, purges, and forced labor camps which spread throughout the socialist homeland. In 1956, however, the Khrushchev report confirmed all that information and, even worse, desacralized the role of Stalin and the CPSU during World War II. The great Chinese famine came abroad as a heterogenesis of the Tiananmen Square Spring. Chen Yizi—former advisor to prime minister and general secretary of the CPC, Zhao Ziyang (fired soon after the 1989 Spring)—brought to the Unites States an internal document from a Chinese institute responsible for reforms in which is said that, between 1949 and 1976 (year of Mao's death), 80 million people died of "unnatural causes"—of those, 43 million died of starvation caused by the "great leap forward."[77]

A review of the Maoist period indicates a doubly disastrous result: on the one hand, forced industrialization and the irreversible proletarianization processes (primitive accumulation based on a very violent domination system) did not even work out; on the other, escaping failure through the post-Maoist reforms meant a drastic and further integration within the dominant economic rationality.[78]

In the 1980s, the Chinese economy was intersected by an incredible movement of private appropriation, by the Communist Party leaders, by the capital accumulated over 20 years of violent and forced industrialization. The loot—which inspired the Soviet leaders in such a degree that they repeated the deed in the 1990s—unfolded along two axes: on the one hand, Deng's reforms[79] introduced a relative freedom, promoting the development which led China to be one of the world's largest economies; on the other, this was accomplished by crushing the lives of factory workers, farmers and the general population. This was summarized by Billeter in four points:

> (1) factory workers were thrown in a brutal competition with the rural exodus of over 100 million peasants, all poor and proletarianized under the Maoist regime; (2) factory workers and peasants which had recently arrived to the areas of intensive industrialization had to live in precarious conditions due to strict control of the organization forms and the internal passport policy; (3) a growth regime marked by a strict control of the media and the repression of all forms of autonomous organization resulted in an extremely high ecological cost, especially in urban areas, which present the highest levels of pollution in the world; (4) while economic indicators show a consistent improvement, the overall quality of life (education, public health, and metropolitan system conditions) was systematically deteriorated.[80]

Thus, since 1949, Chinese society has undergone something like a double seismic event: the violent vertical tremors of the communist regime undermined what the horizontal tremors of market economy eventually brought down altogether. Both seismic events converged in the same trap: communism and market emerged as two variants of the same capitalist rationality, with the communist variant being even more closed and overwhelming than the capitalist one. Indeed, the reforms following Mao's death, during the 1980s, were shattered again in 1989 with the violent repression of the Tiananmen Square movement for democracy.[81] Here is what the Chinese regime and the Latin American "progressive" movements did until they could no longer sustain it and their scheme collapsed (in the second half of the 2010s): they exclusively favored the progression of the "economic reason" in order to convert their power regime (since 1949 in China and since much later in Latin America, although intertwined with the old neo-slave system) into a "modern economic domination." In China's case, the dream is to conceal the power of an economic oligarchy behind the representative politics spectacle. When it comes to Brazil, the "progressive" government tried the neo-developmentalist variant—that is, the Chinese way—at the exact same time China was trying to leave it and managed to link the crises of the political spectacle with the affirmation of the economic rationality as recession.

We are before or within the construction of narratives, and it is hard to accept that China is this industrious *outside* proposed by Arrighi and Sugihara as well as this *inside* represented by the silent transformation concept as introduced by Jullien. In fact, it is not easy to think of the opening of China without the maintenance of the rigid discipline of the Communist Party and the generalized organization of forced labor and its labor camps.

Let us move now to the third point of our book: the crises of globalization and the debate on otherness and the Empire Outside. After the analysis on the exhaustion of the Pink Tide on Latin American politics and about the Kingdom of Middle (China) as a new "old" outside, we will discuss the financial dimensions of contemporary capitalism. Here, we remind that, as Peter Sloterdijk wrote, "the primary fact of the Modern Age was not that the earth goes around the sun, but that money goes around the earth."[82]

NOTES

1. Jean François Billeter, *Contre François Jullien* (Paris: Allia, 2006).
2. Baptiste Mèlés, "Experience and Subjectivity: François Jullien and Jean François Billeter," *Personality and Subjectivity, East and West*, Dec 2009, Clermont-Ferrand, France.
3. François Jullien, *Les Transformations Silencieuses (Chantiers, I)* (Paris: Grasset-Poche, 2009), 32–33. Original emphasis.
4. Ibid., 35.
5. *Laozi*, § 14 *apud* Ibid., 37.
6. Ibid. Emphasis added.
7. Ibid., 38.
8. In 1960, after two revolutions and two world wars, philosopher Emmanuel Levinas is another one who declares Eastern "radical otherness," this time represented by China, which was interestingly in conflict with the Soviet Union at the time: "By abandoning the West, doesn't Russia fear to drown in an Asian civilization, which will also linger behind the structural concrete of dialectics? The yellow peril! It is not racial, it is spiritual. It does not involve inferior values; it involves a radical strangeness, a stranger to the weight of its *past*." *Cf.* Emmanuel Levinas, "Le débat russo-chinois et la dialectique," *Esprit*, no. 10 (1960): 1622–24.
9. François Jullien, *Procès ou création: Une introduction à la pensée des lettrés chinois* (Paris: Éditions Du Seuil, 1989).
10. Edward W. Said, *Orientalism* (New York: Pantheon, 1978).
11. Likewise, during the First World War, Max Weber considered Tsarist Russia the menacing otherness. In a speech given in Nuremberg on August 1, 1916, he said: "A Russian farmer who owns ten hectares of land, an area that would be enough for a German farmer, believes he will starve to death because he is a *savage* who does not know how to use modern agricultural tools." And he then added: "we must have guarantees that the *barbarian* Russians will not return to Eastern Prussia." *Cf.* Max Weber, "Au seil de la troisième année de guerre," in *Discours de guerre et d'après-guerre*, ed. Hinnerk Bruhns, trans. Ostiane Coureau and Pierre de Larminat (Paris: Editions EHESS, 2015) 79, 86.
12. For Jullien, the ancient Chinese thought would be free from the Western transcendent concepts of God and Subject, focusing on the notion of a lively matter in continuous flow, which closely resembles the hylozoism from European philosophers such as the Epicureans, Giordano Bruno, Henri Bersgon, or Gilles Deleuze.
13. Billeter, *Contre François Jullien*, 78–80.

14. Ibid., 13–16.
15. Ibid., 44.
16. François Jullien, *Chemin Faisant: connaître la Chine, relancer la philosophie. Réplique à* *** (Paris: Seuil, 2007).
17. As the Foucaultian aphorism "penser autrement" expresses. See Michel Foucault, "Le Philosophe masqué," in Dits et écrits II, 1976–1988 (Bedroom, Paris: Gallimard, 2001 [1994]), 923–929. The method invoked by François Jullien also resonates with the Russian formalism from the early 20th century in terms of the pragmatism of an ongoing defamiliarization (ostranenie) and, of course, regarding the (immanent) "thought from Outside" theoretically developed by Maurice Blanchot or Gilles Deleuze.
18. François Jullien, *The Book of Beginnings*, trans. Jody Glasdding (New Haven, CT: Yale University Press, 2015), ix.
19. Ibid., 1.
20. Jullien, *Les Transformations Silencieuses*, 79.
21. Fiori, "Sobre o desenvolvimento chinês—II."
22. Ibid.
23. Alain Badiou *apud* Žižek, introduction, 7.
24. Aglietta and Bai, *China's Development*, 33.
25. Ibid., 35.
26. Ibid., 55.
27. Ibid., 83.
28. Here, we may recall the dialectical inversion of Chinese negative otherness presented by Immanuel Levinas. In his article on the Russo-Chinese conflict, he praises Chinese criticism by stigmatizing the Soviet support to merely nationalist young states: "One had to be a bit Chinese to invoke again a cat and recognize the shadow of national-socialism in nationalist anti-capitalisms." *Cf.* Levinas, *Le débat russo*.
29. José Luís Fiori, "Dúvidas e certezas americanas," *Valor Econômico*, May 5, 2015.
30. Jullien, *Les Transformations Silencieuses*, 88.
31. Ibid., 119–20.
32. Ibid.
33. Ibid., 129. Emphasis added.
34. Ibid.
35. Ibid., 150–51.
36. According to Elvin *apud* Jack Goody, *The Theft of History* (Cambridge: Cambridge University Press, 2006), 153.
37. Billeter continues, "Following that same line, Jullien speaks of the Chinese thinkers as if they had always agreed with one another for the simple fact that they were Chinese, as if their thought had never involved aporias, illusions, unacknowledged ambitions, self-serving lies, or a cold determination for taming the spirits, doubts, and lucidity." *Cf.* Billeter, *Contre François Jullien*, 41.
38. Ibid., 32.
39. Sonny Lo, *Hong Kong's Indigenous Democracy: Origins, Evolution and Contentions*, Basingstoke: Palgrave Macmillan, 2015.
40. See the French translation of Goody's *Theft of History*: *Le vol de l'histoire* (Paris: Gallimard/Folio, 2010), 305.
41. Ibid.
42. Joseph Needham, *La Science chinoise et l'Occident (Le grand titrage),* trans. Eugène Jacob (Paris: Seuil, 1973). Originally published in 1969.
43. David Landes, *The Rise of Capitalism* (New York: Macmillan, 1966) 7. *Apud* Nathan Rosenberg and L. E. Birdzell, *How the West Grew Rich*, trans. André Charpentier (New York: Basic Books, 1986).
44. See the French translation of Rosenberg and Birdzell's *How the West Grew Rich: Comment l'occident s'est enrichi* (Paris: Fayard, 1989), 154.
45. Ibid., 155.
46. Ibid., 159.
47. Ibid., 51.

48. Jack Goody, *The Theft of History* (Cambridge: Cambridge University Press, 2006). French trans by Fabienne Durand Bogaert. *Le vol de l'histoire* (Paris: Gallimard/Folio, 2010) *Le Vol de l'histoire*, 147.

49. Jack Goody, *L'Orient en Occident*, trans. Pierre-Anotine Fabre (Cambridge: Seuil, 1999), 277. Originally published in English: *The East in the West* (Cambridge: Cambridge University Press, 1996).

50. Ibid.

51. Ibid., 76.

52. Jack Goody, *Le vol de l'histoire*, 5.

53. Ibid., 151.

54. Ibid., 292.

55. Henry Kissinger, *On China* (New York: Penguin Press, 2011), 531, 408. For a thorough discussion on the subject, see the whole chapter 15.

56. Žižek, Introduction to *Mao: On Practice and Contradiction*, by Mao Tse-Tung (London-New York: Verso, 2007).

57. Ibid., 1.

58. Ibid., 3.

59. Ibid., 10.

60. Ibid., 17.

61. Ibid., 21.

62. Ibid., 25.

63. Ibid., 26.

64. In the left-wing orthodox tradition, even among the most renewing thought, we can always find a reference to the work of Deleuze and Guattari as part of the capitalist sanction regarding the struggles. In the introduction to the English translation of *La Leçon d'Althusser* (1974), for instance, Jacques Rancière, speaking of the 1970s, writes: "This same period saw the flourishing of calls to party and to liberate desire, for which some, too hastily, thought they found the formula in Deleuze and Guattari's 'desiring machines.' Not much later, we saw the 'new philosophy,' which read all of revolutionary history as the appetite for the power of master thinkers unleash its offensive." *Cf.* Jacques Rancière, "Foreword to the english edition," in *Althusser's Lesson*, trans. Emiliano Battista (New York: Continuum, 2011), xv.

65. Mao Tse-Tung *apud* Žižek, introduction, 11. Emphasis added.

66. Žižek, introduction, 11.

67. Mao Tse-Tung, "Talks on questions of philosophy (1964)," in *Mao: On Practice and Contradiction* (London-New York: Verso, 2007), 172–73.

68. *Les aventures de la dialectique*, (Paris: Gallimard-Folio, p. 311).

69. Emannuel Levinas, "Sur l'esprit de Génève," *Esprit*, no. 1 (1956): 96–98. Also found in Levinas, *Les imprévus de l'histoire* (Paris: Fata Morgana, 1994), 142, 144.

70. See Yann Moulier Boutang, *Le Salariat bridé* (Paris: PUF, 1999).

71. Massumi *apud* Žižek, *Organs without Bodies* (New York: Routledge, 2016), 25.

72. Jean-François Billeter, *Chine Trois Fois Muette* (Paris: Allia, 2000), 52. Billeter bases his analysis on the reading of Houa Lin-Chan (Hua Linchan), *Les années rouges* (Paris: Seuil, 1987).

73. Jacques Rancière, *Althusser's Lesson* (New York: Continuum, 2011).

74. Rancière then continues: "In the intervening years, history has taught us not only the limits of the autonomous capacity for initiative attributable to the actors of the Cultural Revolution, it has also revealed the penitentiary realities that accompanied the theses about the re-education of intellectuals through manual labour which, at that time, seemed so consonant with some Western critiques of the division of labour." *Cf.* Rancière, "Foreword to the english edition," xvii.

75. Jasper Becker, *Hungry Ghost: China's Secret Famine* (London: John Murray Publishers, 1996). See also Frank Dikötter, *Mao's Great Famine*. The history of China's most devasting catastrophe, 1958–1962, (London: Bloombury, 2010).

76. Žižek quotes Jung Chang and Jon Halliday in *Mao: The Unknown Story* (New York, Knopf, 2005) and makes a point to remark: "However, not everything in this work should be

taken at face value." See the critique by Andrew Nathan in "Jade and Plastic," *London Review of Books* 27, no. 22 (November 17, 2005): 10–13.

77. According to the same source, 5 million people died in labor camps and 1 million during the Cultural Revolution. See "Chine: le maoïsme aurait cause beacoup plus de morts 'non naturelles' qu'on ne pensait," *Le Monde*, July 20, 1994. Lucien Bianco dedicates a whole chapter to the famine caused by regimes of real socialism, using a considerable amount of sources. Cf. Bianco, *La récidive: Révolution russe, révolution chinoise* (Paris: Gallimard, 2014). See pages 202–221 for the discussion regarding China.

78. Andrew Ross presents an interesting angle on the legacy of Maoism in Europe. According to him, the imperatives of class reeducation and self-modelling on behalf of the personal and collective revolution were reconfigured along the capitalist restoration of the 1970s, helping to create neoliberal subjectivity. Thus, the continuous concern with the self, the modelling of the enterprise-man and enterprise-society, would have an antecedent in the Cultural Revolution of the 1960s. This is a more interesting approach to subjectivity related to China than the mere accusation of a supposed ideology hiding behind an exoticizing sinology. See Andrew Ross, "China's next cultural revolution?" in *Nice Work If You Can Get It: Life and Labor in Precarious Times* (New York Univ. Press: New York-London, 2009).

79. It is important to mention that, as Lucien Bianco points out, Deng updated Boukharine's motto: "Get rich!" *Cf.* Bianco, *La récidive*, 47.

80. Ibid., 56.

81. Twenty-seven years later, "[t]he families of those killed during the 1989 Tiananmen crackdown have accused Beijing of subjecting them to nearly three decades of 'white terror' in a bid to stop them speaking out about the massacre. In an open letter, published on Wednesday ahead of the 27th anniversary of the protests, the Tiananmen Mothers campaigning group said its members had been spied on, detained and threatened by security agents as part of attempts to cover up the killings." *Cf.* Tom Phillips, "Families of Tiananmen Square victims accuse Beijing of three decades of 'white terror,'" *The Guardian*, June 1, 2016.

82. Peter Sloterdijk, *In the World Interior of Capital* (Cambridge: Polity Press, 2013), p. 9.

Chapter Five

The Schizocene of Money

Meanwhile the role played by the financial sector continues to be regarded as the fundamental problem of contemporary capitalism. For some, this is a case of a "deviation," of a fictitious and unreal sphere (as opposed to the sphere of material—industrial—economy), while for others, it is a case of one of the moral characteristics of Western civilization: the infinite debt. As anthropologist David Graeber stresses: "If one looks at the history of debt, then, what one discovers first of all is profound *moral* confusion."[1] In this chapter, we thus propose to weigh this impasse, which, as seen from South America, appears as a double and dramatic crisis: a crisis of globalization as well as a crisis of the attempt to step outside of globalization. As such, the exhaustion of governments and of discourses that were considered "progressive" in South America appears to put back the subcontinent in its historical and geographic "prison." In the crude terms used by mainstream geopolitics: "Academics and journalists are fond of writing that the continent is 'at a crossroads'—as in about to embark at last on its great future. We would argue that, geographically speaking, it is less a crossroads than at the *bottom* of the world; there's a lot going on all over this vast space, but the problem is that much of it is going on a long way from anywhere other than itself."[2] But it is precisely this perspective, this "bottom" that is of interest to us.

BETWEEN CHINA AND THE WEST: *FIAT* MONEY

The debates on "inside" and "outside," as well as those on the Needham and the Talcot Parsons paradox,[3] reappear as we discuss the workings of finance, its origins, and its relations with currency and money in general. Max Weber, Werner Sombart, and Joseph Schumpeter have assigned a fundamental role to the invention of rational or scientific accounting in the development of capitalism. Resuming those analyses, Jack Goody reminds us that the invention of the double-entry technique, in the Italian Renaissance, was the fundamental mechanism for the activities of Italian bankers to establish an almost absolute monopoly over the emerging European finance, leading the banks of Venice, Florence, and Genoa to dominate the money markets of London and Bruges.[4] David Orrell and Roman Chlupaty wrote that "the technique got its name from the fact that every transaction was entered in two different accounts, once as a debit and once as a credit. Thus an asset in one account appeared as a liability in another. The method helped detect errors since the sum of credits over all accounts should be balanced by the sum of debits, as positives are balanced by negatives."[5] Goody highlights that this was not just a technique, but a turning point, especially from the social and cognitive point of view. Since the Sumerian times, "the adoption of money was part of a generalized shift toward the dominance of calculation in our lives."[6] The innovation, which would end up giving the traders of Italian independent cities the conditions to dominate the emerging monetary circuits, consisted in this method of writing the exchange transactions between different parts of a single company or between different accounts.[7] Merchants who could not read or write were left behind. No wonder the double-entry bookkeeping was first systematized by a mathematician who was, at the same time, a collaborator with Leonardo da Vinci and a part-time magician.[8] In this sense, bookkeeping is contemporary to the popularization of the press, a new graphic reason[9] that allowed knowledge to raise a greater degree of complexity and abstraction. Together, those changes fostered new monetary practices, a real transformation at the heart of currency operations, which would become one of the conditions for the development of capitalism.

The fiction of treating an amount of deposited money as credit (customer account) and debit (from the bank vis-à-vis the customer) allowed to calculate profits and losses at all times, without having to stop the flow: the technique "gave a quick snapshot of profitability, with enduring consequences for our view of the world."[10] This meant an innovation: individual accounts (at the micro level) would merge with social accounts (corporate, as Ernest Kantorowicz[11] would say), forming a continuous effluence[12] in which they commingle with one another in a common flow.[13] At that moment, the ability to generate an aggregate effect from multiple bookkeeping practices in different countries, with the most diverse origins, producing a superpersonal mass of credits and debts, allowed the Italian bankers to escape the eyes and the grip of feudal, national, or religious authorities. Like a slippery fish, money was gradually breaking away from the intricate coding grids, from the limitations, barriers, bureaucracies, and regulations imposed from above. The fluidization of money is contemporary to its increasingly free spill over new social and mercantile circuits, making them brew, producing new connections, fostering new assemblages. This socialization provided by transversal monetization represents a mutation of the very nature of money. At the same time, the powers being left behind do not fail to react to this change. They see the decoded flows of money with the terror of those who witness their own dissolving death. And so starts the long history of denouncing this "free" currency—increasingly landless, stateless, and atheist "free" currency, an active force of nihilism that terrorizes old social structures, making them react to the threats of their dissolution.[14]

Amid the intellectual effervescence of the 1960s, the French monetary economist Bernard Schmitt wrote: "it was money (that) allowed the industrial revolutions, thus raising the conditions for the full realization of the very money."[15] In slightly different terms, John Maynard Keynes also said that money is first and constitutive regarding capitalism, and that the latter would not have been possible without the prior monetization of society. According to him, the bifurcation which determines the birth of capitalist modernity occurred with the significant increase of money accumulation enabled by the monetary innovations. The first "innovator," as mentioned by Keynes, would be the English privateer Francis Drake, who in 1589 plundered the gold which the Spaniards had looted from the Incas and Aztecs.[16] In fact, the original peoples from the American continent supplied the Old World economy with precious metals for centuries after the European invasion. This bullionist influx structured the workings of mercantile monetary economy based on the reason of state and the overcoding of flows made possible by the colonial loot, the continental wars, and the slave labor in gold and silver mines.[17]

However, the decisive innovation for capitalist development was not the quantitative multiplication of money, nor even the banking unfolding of credit, which existed for thousands of years among the Sumerians. The decisive innovation for capitalism occurred with the qualitative transformation of money, the moment it reached its most abstract form, resulting in its change of nature. We refer here to Fiat Money, to its launch on a solely imagined return—that is, its creation *ex nihilo*, exempted from material correspondence in some kind of backing or counter deposits. J. K. Galbraith opens up his book *Money: Whence It Came, Where It Went* stating that "money is a very old instrument, but the idea that this is a trustworthy artefact which would be acceptable without close scrutiny or objections is . . . a circumstance of the last century."[18] This transition could only occur due to a profound transformation in the nature of fiat money—better yet, in the very notion of *trust*. Money is no longer a commodity, something exchangeable and commensurable, but a "representation of debts and credits"[19] that gains autonomy from its support and is set to flow as a sign of an abstract power based on trust. Despite somewhat schematic and sometimes apologetic,[20] François Rachline radicalizes this approach and identifies abstract money as the deflagrating agent of both European modernity and its alternative paths.

It marks the exceptionality of the West in the history of capitalism. In other words, since the European Renaissance, money managed to deterritorialize itself in relation to sovereign, feudal, and religious powers—a historical process which did not occur in China or India.

According to Rachline, echoed by Rosenberg and and Birdzell, the fragmentation of empires presents a favorable circumstance for the advance of payment and bookkeeping techniques, which merchant and banker communities have developed and practiced *despite* the centralizing powers of states and sovereigns—sometimes *against* those powers. They had to circumvent hierarchical and bureaucratic codification, evade the vigilant discipline against monetary flows which eluded the state grip, and escape the eager pursuit of regulations and taxation as if, in the face of despotic asphyxiation, monetary flows had to blur the sovereign image imprinted in coins. More than that, money had to be stripped from the indexation of standards or terms of comparison—the *royal measure* which currency should always refer to—that never ceases to subject it to the sovereign's accounts, to the palace spreadsheets involving metaphysical and political transcendence. At first, state powers adopted the *échangiste*[21] approach to money since they needed the equivalence principle to keep it under control within social circuits while they established themselves as the only non-equivalent instance, the transcendent overseer and gauge which imposed an underlying verticalization on the whole economy, an *overcoding* from above. Until the European Renaissance, the sovereign or feudal lord ultimately referred money to himself, as if value emanated from his own transcendent condition, a quasi-cause of his value. From the perspective of sovereignty over territories, money had to remain within the circuit between taxes and the capture of the surplus labor by the state, so all economic activities could be taxed and the surplus addressed to the state vertex.

From this point of view, the pyramidal hierarchy of the Chinese Celestial Bureaucracy—praised by authors like the historian Giovanni Arrighi or the philosopher François Jullien as we have seen above in chapters 4 and 5—played a negative role in the field of monetary innovations. The imperial power promoted the monetization of society, essential to value extraction through taxes, while also warding off its decodification, preventing "free" flows from escaping. From the celestial highs, the sovereign had to keep the country's economy on a short leash, because he believed everything *should* converge. In that sense, Billeter (as well as Wladimir Pomar) is right when he criticizes how the "industrious revolution," theorized by Kaoro Sugihara and resumed by Arrighi,[22] still remained under the rule of a limited and traditionalist imperial sovereignty, wrapped in a static temporality and organized in a stratified way: in other words, a self-sufficiency economy closer to feudalism, whose reverence from Western philosophers and economists has no other explanation than an urgent need to identify a shortcut to some Outside of the capitalist industrial revolution, which shaped globalization. According to the Franco-German sinologist Étienne Balazs,

> Chinese skill and inventiveness, which have provided so many of the amenities of life, such as silk, tea, porcelain, paper, and printing, would no doubt have continued to enrich China and would probably have brought it to the threshold of the industrial age, if they had not been stifled by state control. It was the state that killed technological invention in China. Not only in the sense that it crushed *ab ovo* anything that went against or gave the appearance of going against its own interests, but also because of the customary attitudes so firmly implanted for reasons of state. An atmosphere of routine, traditionalism, and immobility, in which any innovation or initiative not demanded or sanctioned in advance is regarded with suspicion.[23]

Leaning on Balazs, Deleuze and Guattari conclude: "When Étienne Balazs asks why capitalism wasn't born in China in the thirteenth century, when all the necessary scientific and technical conditions nevertheless seemed to be present, the answer lies in the State, which closed the mines as soon as the reserves of metal were judged sufficient, and which retained a monopoly or a narrow control over commerce (the merchant as functionary)."[24]

In China, even if other conditions were met, the implementation of a paramount condition was interdicted: the decoding of monetary flows, which was deliberately blocked through the rigid control that the sovereign power exerted over money and its innovations. Therefore, money as means of payment did not convert into pure sign, as it had happened in the city-states of the Mediterranean Sea. One can also find such political interdiction to the development of capitalism in the Roman Empire.[25] Moreover, it is worth rethinking the implosion of the Soviet Empire in the light of this theological-state blockage. In *Anti-Oedipus*, in the wake of Karl Wittfogel's studies, Deleuze and Guattari relate the logic of the Stalinist USSR to the "Asian model"—that is, a despotic machine obsessively committed to overcoding flows, a state machine hooked on the latency of terror.[26] According to the authors, the difference between the West and the USSR in terms of how capitalism worked did not exactly lie in a greater or lesser presence of the state, since one and the other operated a modality of state capitalism, but in the *type of coexistence* of both the capital and the state machine. In Keynesian Fordism, axioms would penetrate the capitalist machine so that currency could function with greater fluidity, socializing itself to constitute the social or welfare state; meanwhile, in the "Asian" USSR, state capitalism remained saturated with imperial and bureaucratic territorializations and their paranoid overloads. While the coexistence between state and capital was *intensive* in the West, where capital put the states to work according to its immanent model of governing flows, in the USSR, this coexistence remained *extensive*—that is, the state sought to settle in an Outside, which jammed the currency machine. Similarly, between the 15th and 18th centuries, the Chinese imperial government, trapped by a material stock logic, kept the merchant class confined and systematically thwarted all efforts of deterritorialization by sea, including having the bulk of the commercial fleet burn. Against the liberation of flows, there was a labyrinth of regulations regarding mercantile and banking activities, to the point that money could only be used in the strict forms set by the emperor.[27] In this respect, it is worth noting the historical milestone of 1429, when Zhu Gaochi, the fourth emperor of the Ming dynasty, decided to abolish sea expeditions, drastically reduce the merchant fleet, and ban their armaments,[28] in a paranoid reaction to the deterritorializing lines of the sea, which opened market flows, unblocking the monetary route. Even if it refers to a different kind of flow, Trump's Muslim Ban just at the beginning of his presidency is a similar paranoid decision, deepening US decline.

Thus, if the mutation of currency in terms of transversal socialization is a condition for the existence of capitalism,[29] then the history of monetary inventions was the history of different attempts at escaping, rupturing, and feinting, which managed—*against the state*, against its ongoing effort to hinder decoded flows—to break the absolutist coding of money.[30] By reaching critical mass outside those conditions, money got rid of traditional codes and territorialities to exist as pure sing, both abstract and centerless. Decoded, it only exists in a state of perpetual movement, in a continuous spill that constitutes new mercantile, productive, and social circuits. The emergence of capitalism occurs at the moment the "free" flows of currency combine with the "free" flows of labor, which, in turn, derive from enclosures, colonization, bloody wars, violent population displacements, and other processes of "liberating violence." At that moment, we witness the consolidation of a dualism at the core of money, which becomes *two*, divided into poles of different natures. The first is the money we carry in our pockets, with which we compare wages and prices; the other is the money-capital, intangible to us and surrounded by the miracle of money creation managed by banks. Between the former—the material money, used as means of payment—and the latter—the abstract money, converted into capital's power—there is a whole interval whose elements are more or less subject to one or another tendency, although neither will ever be experienced in a pure state. Regarding this duality, Bernard Schmitt highlights the difference in nature between the flow of quantifiable money, which measures earnings and consumer goods, and the unmeasurable flow of capital money.[31] The existence of the latter in parallel to the money used in exchanges and payments marks the difference between capitalism and other modes of domination. When it comes to abstract money, the unbridled power of social fermentation is then operated by central banks, public and private banks, as well as by finance as an integrated system, when the political character of capital is directly expressed as monetary policy toward which everything else gravitates. In Schmitt's theory of the two flows, one refers to the means of payment and the circulation of commodities while the other is directly linked to the creation of money by the banks. While representative entities of the working class and the Left in general—still restricted to a distributist, quantitativist, and *échangiste* notion of currency—are more prepared and routinized to dispute the former, capital applies its politics directly over the latter—the field of deterritorialized power that determines the "value of values," the grounds for the creation of money and sign-money.

According to B. Schmitt, money is not the ether of exchanges, a mere medium for facilitating economic transactions that could be removed from operations as if the essence of barter lay behind it. In his recent book about debt, David Graeber dedicates a whole chapter to explain that barter is just a "myth." Money does not come to life in intersubjective relations, but in the very production of subjects. It incorporates the deterritorializing power to social production for the constant revolution of means, techniques, and productive arrangements. This is the liquidity injection rationale, the logic of the invention of flows. In this sense, one condition for the emergence of capitalism has been logically and historically accomplished with the alliance between money and capital, when they begin working together in the constitution of the social.[32] Still, according to B. Schmitt, capitalism therefore implies a profound and structuring dissimulation[33] when it puts the particular variation of each flow (wage/purchase power = ∂x and currency/capital = ∂y) in relation to each other.[34] This gives rise to a differential function and not just a simple difference. What matters is not exactly a relational flow, but the differential relation between flows, which is productivity to the nth degree. The logic is relational *and* productive: it does not boil down to two preexisting terms that, once in a relationship, vary; instead, both are actually variations that relate to each other, which causes this relationship to be fluently productive. Thus, the surplus value is extracted from the differential element $\partial x / \partial y$, a volumetric function, which ultimately results in a torrent of surplus value.[35] This immense discharge gathered throughout the surface of capital can no longer be measured or captured by any quantitative term of comparison; neither can it be understood by the classical theory of value, by the simple deduction between necessary labor and surplus labor, such as in the calculations of David Ricardo's political economy.[36] In capitalism, credit-money is fully integrated to the real economy, being inseparable from the productive circles, where banks play an inescapable role. All the social productive logic is linked to the creation of money *ex nihilo*, with which the bank, so to speak, electrifies money with a charge of trust to magnetize the agents and productive assemblages. A theory of value in the Ricardian style could only work under such conditions that the value could be calculated by a simple difference between quantities—say, between the quantum of time regarding the labor performed by the worker and the quantum of labor appropriated by the employer. This is very far from happening under capitalist conditions. When the exploitation relationship becomes a differential function, the difference in quantitative terms completely loses relevance and becomes just a particular case in the face of the surplus value flow that money-capital collects from the entire social fabric charged with trust—something similar to which Marx theorizes in the *Grundrisse* as the immediate productive power of the General Intellect.[37] In capitalism, financial econometrics is what better grasps and allows the further development of this differential equation between monetary flows according to a system of variable quotients of production and demand projected on the future—that is to say, on an economy of the possible. Therefore, in capitalism, the whole market works as a future market under the hegemony of money as pure sign.[38]

Based on the baroque and counterintuitive theory of B. Schmitt, Deleuze and Guattari devised a schizophrenic theory of money in *Anti-Oedipus*.[39] Inasmuch as the complete monetization of society is a condition of capitalism, the very desire that drives the social dynamics becomes fully invested in the monetary flows. With capitalist development, the marked distinction between desire and productive infrastructure disappears. The social logic is immediately guided by currency flows, without the need for a "metaphysical leap" in ideology, hegemony, or other theories of superstructure. The total financialization of life follows this short circuit of the theory of value and all mediations, from the microtextures of everyday life to the governance of globalization by the world financial system. Meanwhile the individual body and its libidinal economy become points for the action of opposing forces, between desiring and social production.[40] According to Deleuze and Guattari, capitalism is the most terrifyingly immanent machine because it controls "from within," from the modulation of bodies, their sensitivities, and the emotions they raise.[41] While monetary flows become the spiritual core of the system, control is then exercised in the continuum between desire and subjectivity, with maximum identity between the *socius* and desire, between currency and body. Hence the exploitation relationship, when it becomes differential, shifts to the interior of the intensive constitution of subjects, as a pre-subjective, molecular and, as Foucault would say, ethical plane.[42] It is no longer a matter of a simple quantitative confrontation of forces between worker and employer; instead, it is now a qualitative relationship in which the "employer" was raised to the condition of total abstraction: that is, a relationship of subjective debt.[43] The virtually absolute liquidity of money-capital implies, in return, the endless repetition of this subjective debt, since—and *pour cause*—the relationship of each one with oneself becomes pregnant with desire and capital at the same time. From those conditions for the existence of capitalism, a fundamental ambiguity emerges in the capitalist machine, for if monetary flows organize the process of capital and, at the same time, molecularly infiltrate the social fabric, then the conditions for a "machinic revolution" arise through the production of another currency. A revolution that does not have to go through the stage of conquering the superstructural mediations and their Winter Palaces, whether this means regaining the state, conquering hegemony, or performing the dialectical inversions of the dictatorship of the proletariat. On the contrary, such archaisms are no more than obstacles reworked by the very capitalist process in order to postpone the schizophrenia of the flows and keep them within workable limits: it is no coincidence we can find them both in communist China and in Putin's post-Soviet Russia. With the right hand, capital deterritorializes flows; with the left one, it reterritorializes them in new-old icons and images. In this sense, according to *Anti-Oedipus*, the state is a neo-archaism within the capitalist system, some sort of avatar of capital reterritorialized in the image of old sovereign powers, yet strategically reapplied to interpose regulations, infiltrate axioms, and raise obstacles to the definitive liberation of desiring/monetary flows, to the advent of the absolute limit which would lead it to eschatological collapse. Therefore, one of the authors' most controversial provocations is not that they oppose the schizophrenic flow of the money-desire in a "revival of the fascist 'economic solution'" (neo-fundamentalist); instead, it may be that they "go in the opposite direction . . . for perhaps the flows are not yet deterritorialized enough, not decoded enough, from the viewpoint of a theory and a practice of a highly schizophrenic character. Not to withdraw from the process, but to go further, to 'accelerate the process'"[44] would take money beyond the evil alliance established between currency, as an abstract machine, and capital, as diagnosed by B. Schmitt.

MONEY AND DESIRE

Indeed, the description of the monetized economy through desiring flows is, to some extent, already present in Keynes, who, according to Deleuze and Guattari, was the first to "reintroduce money into desire."[45] In a controversial article on the ephemeral and unsuccessful return of the gold standard in the United Kingdom, in 1924, Keynes chose a political metaphor to depict such displacement: by means of a "modern alchemy," he says, "*gold*, originally stationed in heaven with his consort silver, as Sun and Moon, having first doffed his *sacred* attributes and come to earth as an autocrat, may next descend to the sober status of a *constitutional king* with a Cabinet of banks."[46] Demoted to a sober status in Keynes's description, the gold standard was definitely abandoned in the United Kingdom in 1931. Once again, the desecration of the role of gold paves the way to paper money and money as pure sign, point of power, the last and final stage of its liberation from despotic signifiers.[47] As Dostoyevsky once put it, "money is coined liberty"[48] It is no wonder that the inventor of macroeconomics employed a profusion of notions of individual and social psychology to depict the monetized economy: propensity to consume, preference for liquidity, optimism/pessimism to endeavor, depression and euphoria, paranoid spirals. The very notion of "effective demand" is linked to an impersonal and aggregated subjectivity obtained by combining the many singularities that make up the dynamic reality of economy. At times it seems that, for Keynes, the role of the government is comparable to the task of a motivational psychologist who must be alert to the country's entrepreneurial and consumer trends, be ready to react to psychic declines, and induce healthy optimism in the economic agents. The formula for macroeconomic calibration developed in the *The General Theory of Employment, Interest, and Money*[49] works through a constant diagnosis of social moods, offering a number of prophylaxes and prescriptions for each case of imbalance in the organism.

"One of the defining features of money is that there is never quite enough."[50] Once developed, the journey of capitalism in the West resulted in conflicting, multidirectional, and contradictory historical processes, but its unfolding eventually gave a certain answer to the money enigma. This answer came with the progressive replacement of the *promised* thing by the *promise* itself—that is to say, we went from the palpable product to the abstract trust placed in the operation of the system as a whole. The mutation of currency was the cause and effect of this change of temporality: no longer restricted to the simultaneity plane of the present—the time par excellence for exchanges, for the circulation among equals, of the finite and payable debts—it is now a transversal component of desire and the *socius*. Exceeding the time of debt resolution in the present, money becomes a volumetric figure, a cone filled with multiple temporalities: past, present, and future integrated according to an economy of the possible in which the present is nothing more than a point among a multitude. This reality does not allow one-dimensional measurements of quantity and can only be parameterized by tendencies, gradients, oscillations, and differential relations. Thus, currency no longer refers to the simultaneity of exchanges; it now directly lies in the becoming, in the perspectives, in the memories, and in the continuous and qualitative modulation of time. The capitalist market absorbs the future and the past, adjusting to the variations of the demand over time in a system of prices with no indexation. When capital consolidates its alliance with the sign-currency—which is *put to work* as money-capital—the "money as capital" that reigns in Marx's D-D' cycle—the confluence of desire and money reaches the threshold of paroxysm. "The appearance, the surging forth of now decoded flows . . . pour over the *socius*, crossing it from one end to the other."[51] Capitalist schizophrenia is marked by the dissemination of the destructive/creative force of its deterritorialization, which dissolves social forms, desecrates values and traditions, and "melts all that is solid into air."[52]

At the same time, it internalizes a subjective debt, preventing desire from reaching its full potential. Desire is then a mitigated desire, plagued by death, which comes from within and connects it to the *socius* (death instinct).[53] Currency captures from this continuous differential, which is both desiring and economic, until it forms the massive liquidity blast of surplus value, the capitalist's Body without Organs.[54] In the most eschatological passage of *Anti-Oedipus*, revolution is detected when the gap produced by capital between money and desire is surpassed, when full accelerationism is attained, so that flows can run free on a deterritorialized BwO.[55] This would mean seeing the Capitalocene through the end, abolishing it, giving way to the Schizocene for a new alliance, a *new currency*.

After the money convertible to gold, there emerged the inconvertible paper money, *fiat* money, whose value rests solely on the trust people place in it.[56] Although some economists, nostalgic of the codification (of territorial, national, or theistic/backed currency), often deride a supposed mystical force intrinsic to the concept of trust, the fact is that the *production of trust* was a necessary condition to the development of capitalist economy, as it enabled the conjunction of decoded flows: on the one hand, the flow of "free" work; on the other, the "free" monetary flow. In other words, both ends of the primitive accumulation process, which has to be constantly reproduced by capital.[57] There is no possible return. The set of innovations, mutations, and historical processes on the basis of capitalism has led trust to become a kind of belief with an immediate political and economic dimension, a belief that was systematized, for instance, in the civic religion which Machiavelli had identified in the city-state of Florence during the 16th century's upheaval or in what Spinoza, who witnessed the banking and mercantile prosperity of the Seven United Provinces (Netherland), called the power of the multitudes. It was no accident that the deterritorialization of currency found a fertile ground in the economic-political chess of the Italian city-states and, later, in the decentralized Netherlands of the 17th century. Only later this conjunction process of social and desiring flows, which currency expresses, would reach a great national power, the British Empire, which established itself as a thalassocracy of planetary scope: an Empire of the sea, to the sea, and from the sea.[58] The state of England, which deterritorialized itself and became a seven-sea empire where the sun never sets, was something like a ship or a fish, as Carl Schmitt described it, the great Leviathan, "the mobile centre of a world empire, the possessions of which were strewn in no coherent pattern over all continents."[59] Therefore, the maximum fluidity, deregulation, and flexibility that finance will acquire in the twentieth century is a development of this *long duration* process throughout cycles of accumulations and (de)territorialization of capitalism—although this is not the approach adopted by historians of world economy like Fernand Braudel, Phillippe Norel, Immanuel Wallerstein, or even Arrighi.[60]

As we have seen, in this melting pot we have the historical-political dispute between the despotic overcoding machine and the decoded flow of money, which spanned many centuries and expressed itself in the debates of the history of economic thought. This dispute will be decided, according to a decisive tendency, toward the second pole. When the weight of state power is circumvented by the Italian mercantile city-state or the Amsterdam Stock Exchange, the diachrony of the very credit causes money to escape the exchange logic; then debt is issued over debt, bond over bond, opening the hereafter to the penetration of capital as well as desire, which is colonized by it. Two lines of circulation assert themselves in those disputes: one—vertical, synchronic, top-down—depends on the power of direct appropriation (looting, taxes, and other kinds of plundering); the other—transversal, diachronic, bottom-up—functions through networks based on the ability to put into circulation what has been produced or to set relations through services.[61] Finance takes root from there, with the rupture of *1)* the *synchronicity* between acts of issuing and extinguishing debts, *2)* the *commensurability* of the creditor and debtor poles, and *3)* the *materiality* of currency backing, imposed by state and feudal codes. The acceleration of time, usually described as a structural compulsion of capitalism, corresponds to a growing expansion in between those two moments, creating a third dimension, a volumetric dimension filled with currents, waves, bubbles, spills, and all sorts of phenomena describable by the fluid mechanics. In the 21st century, every attempt to refer the artifice (currency) to some natural state (by the metal standard, by the spiritual interdiction of usury, by paralyzing taxation) runs the serious risk of getting involved in some kind of "material populism," which not only reinforces the impotence in the face of the deterritorialized power of capital, but also endows with good conscious a depressing mechanism of desire, which Nietzsche identifies with reactive nihilism.

According to Clément Rosset, the "artifice (fiction) becomes artificial only when it disguises itself as nature: it is the desire of belonging to a transcendent principle (nature) that denies the gesture its original authenticity, its true 'natural.'"[62] Thus, "the future takes the place of the past as effect takes the place of cause: the naturalist illusion mirrors the origin of customs in nature when it is the constitution of all nature that derives from customs."[63] The fiction of money dispenses with metal backing and supports its grip by producing trust: "Perhaps as a kind of warning, in 1963, the words 'payable to the bearer on demand' were removed from newly issued dollar notes, to be replaced with 'in God we trust.'"[64] The mutation of sign-money is based on the constitution of an anthropological transformation of values, when trust emerges as the integrated expression of this new systemic unit, which is both qualitative multiplicity (of flows, networks, exodus) and immanent unit of measurement (the incommensurability of the desire that determines the value of values, the convergence of desire and the *socius*). Initially a countable unit under a slow circulation scheme coded by state and territory, sign-money resurfaces as an abstract, deterritorialized, and future-oriented device—a *projection*, as Bernard Schmitt would say.[65] The notion of projection is fundamental to understand the "tensor" nature of *fiat* money: banks can only create money out of nothing, as a simple promise to pay, because it inspires confidence. Thus, creating money is nothing more than the projection of a bank debt into the productive economy.[66] Both the material and immaterial dimensions of money—as objective asset and as differential relation between flows—are intertwined, and one of the challenges is to face the difficulties that this poses at every attempt to understand and democratically assert the dissimulation and ambiguity of money.[67] That is to say, one must grasp the subjective desire which, beyond the capitalist capture, also emanates from money, representing a liberation force.

The monetary flow promotes the endless multiplication of commuting among the operators of the economic activity. Funds are repeatedly used and reused, money turns over, moves, circulates *n* times in a derivative way until it reaches the fermentation which the simple material quantity refuses. This results from the very mercantile activity when it escapes the regulatory anchors: the merchant buys to sell and sells to buy, and so he needs credit in advance. The merchant, as the worker and the entrepreneur, is a debtor in structural terms, since he depends on the credit-money with which he gives consistency to the capital turnover. The creation *ex nihilo*, as a result of the circulation acceleration, as well as the folding and unfolding of monetary circuits, is not essentially based on the looting of nature, but on the intensification of social relations.

What matters here is grasping what grants volume to the *in between* which the transversal circulation of currency produces, generating trust, and, at the same time, empties it of content, causing loss of trust and the bursting of bubbles. How does this trust in the future develop to the point of becoming sound belief? And how is it possible that currency expresses this trust to the point of spreading a live movement beyond its origins as credit or tax? First of all, we must assume the notion of trust in its ontological context, despite the mystifications produced by the endless diatribes going on between orthodox and heterodox economists. The challenge is not discarding this terrain of trust even though this is a central and often mystified notion among orthodox economists. Instead of being trapped by the constitutive dualism of money, through which the process of capital works, we should develop the content *in between*—the many overlapping strata between trust and things to come, between monetary and desiring flows. Simply identifying the capitalist schizophrenic logic of currency is not enough to solve the ambivalences and paradoxes that multiply within the political and organizational field of the struggles. On the one hand, state sovereign power is the condition for a forceful circulation according to vertical lines which obsessively overcode the flows. On the other, circulation feeds itself from trust, which it confirms by transversal lines that overflow on all sides and change in qualitative terms. Ultimately, none of those two aspects of circulation completely prevail. To a certain degree, they keep feeding on and interfering with each another. While the vertical circulation of striated currency (marked by exchanges, means of payment, commensurability, and equivalence principle) is bound to the sovereign power and, as such, exists in an intersubjective and contractual dimension of the Hobbesian kind, the other, the circulation of flat currency as pure sign (characterized by the power of capital, creation *ex nihilo*, incommensurability, and lack of equivalence measure), implies the transversal mobilization that feeds on trust and, thus, on a continuous variation, which does not fit into a model of contract between parties. As we have seen in Bernard Schmitt's work, the process of capital is essentially based on the extraction of surplus value from the relation between two variations—that is, the wage variation in relation to prices and the variation of investments created *ex nihilo* against the reflux deferred through time (which establishes it as permanent income)—resulting in the enormous flow of machinc surplus value. It is as if the dynamics of absolute (human) and relative (machinic) surplus value described by Marx in *Capital* were articulated in a differential relation. Regarding the issue of currency systems, this implies the difference between the *ex*tension of money quantitatively appropriated, exchanged, or plundered and the *in*tension of money qualitatively captured from the socio-machinic productivity.

One must understand the socializing tension induced by sign-money, which allows capital to work in extension and depth, whose encompassing power is such that nothing can be left out.[68] According to Orrell and Chlupaty, both natures of money have never stopped walking hand in hand: "If the U.S. dollar, which was still the main reserve currency for the global economy, was not backed by gold, then what was it backed by?" And they remember that one answer "was to say it was backed with oil instead of gold." Another was to say that "dollars were backed by military force."[69] From Rachline's point of view, this is partially solved by the difference between Hobbes and Locke regarding their notions of contract: according to the former, contracts are irreversible and cannot be broken, while the latter believes the community is allowed to break them. The fact is that, in order to go beyond the duality of currency, we must think of non-contractual relations. That is to say, we must approach this in terms of subjectivity production instead of inter-subjectively developed relations—a gap Deleuze and Guattari sought to fill: first, with the widespread machinism in *Anti-Oedipus* (1972); later, in *Kafka: Toward a Minor Literature* (1975), with the concept of "*agencement*" (a term we prefer to the English translation, "assemblage"). Marx's "Fragment on Machines" as well as Georg Simmel's *Philosophie des Geldes* had already led us in that direction.[70] For the latter, currency is a specific form of credit, insofar as the debtor entity is neither an individual nor even a legal person, but an impersonal and trans-subjective entity based on the merchant community in its multiple connections, disjunctions, and combinations. Between debtor and creditor, there is a third instance, different in nature, consisting of the whole social body, which bears the power to sustain value. There, the subjective—or, we could say, common—nature of currency is already present.[71] "This is why," André Orléan says, "the logic of contract does not apply: money is a claim whose force derives form a tacit, provisional interest among the community's members, and not from some formal undertaking that constraints all of them to accept it under any and all circumstances."[72] It is the power of the multitude and the kind of trust it produces that sustains the monetary circulation. This very monetary circulation, in turn, fosters trust. According to Deleuze and Guattari, this machinic connective tissue, whose integration is carried out by banks, is precisely what the process of capital encompasses. Indeed, Simmel splits the notion of trust in two. Besides a more mundane trust—which develops from beliefs or objective knowledge that, based on the observation of past events, engender representations of what is to come—there would be a second type of trust that we could call credit. The first one, based on memory, is of the inductive kind. The other, an almost religious type of trust, is more related to belief—here we would almost dare saying "faith"—than to knowledge. In this case, the self abandons itself without much resistance, as in some sort of second innocence.[73]

During the 18th century, London could finally accomplish what the Republic of Venice had previously drafted: to consolidate the long-term debt, transforming the overall structure of production in a sustainable way. Until then, all attempts at financial invention had bumped into an opposing sovereign power or an antagonistic network of state bodies that worked as a systemic blockage. The British Empire case is the first one where the very state was fully subjected to the power of money, becoming part of the immanent operating model of capitalism. The state acts to guarantee the conditions for deregulation, for free trade. The transversality of "free" markets is in tune with the mitigated verticality of the British monarchy. Impaired by successive revolutions and riots, many of which were born at sea,[74] the English royalty abides by the market. The sovereign bows down to the monetary flows and, at the risk of deposition, is forced to fulfill the obligation of a safe and sound return.[75] As Keynes put it, gold had to divest itself of the sacred aura to preside over a parliament of bankers, symbolically assembled under the decorative figure of a constitutional monarch. The real monarch, however, is not in the heights, but on every pound sterling, under the rule of banks and stock exchanges. Thus we can think of capitalist development as the history of this dualism of money up to the point when state resistance gave way altogether. Taking its first steps at the Italian city-states of the Renaissance, where Machiavelli lived, and then going through the economic and financial effervescence of Spinoza's Netherlands, sign-money finally consolidated into a new type of institution, the Bank of England (1964), the mother of all Central Banks. The alternative modernity of insurgent power[76] was traversed by the creation of sign-money.

The Central Bank is the institution that consolidates the policy of capital as monetary politics. It results from the long struggle which ends up with the power of money supplanting the state overcoding. Through the Central Bank the state machine is employed as an instrument of capitalist production for the first time. Now, sign-money, free from serving the modern theory of representation, is entirely devoted to sustaining the huge deterritorialized and impersonal flow of desire. The Bank of England worked as a model for all subsequent central banks, standing at the hinge between two types of money as the pivot of the dualism of monetary flows.[77] Therefore, monetary politics will be first in capitalism, transmutating the flows and putting the differential equation of surplus value to work. It is the capitalist solution for the dilemma of power (the money of the Prince) and the common (the money of the People), between the "two bodies of the Central Bank."[78] Thus the dual dimension of the individual and the legal entities, one corresponding to the natural body and the other embodied in the figure of the King: "Two bodies but one person, because the king as King was 'incorporated with his subjects, and they with him.'"[79] But the amalgamation of one and the other occurred in a paradoxical way: the mortal figure of the King was God-Made while the Immortal King was Man-Made.[80] Kantorowicz quotes Plowden to further his thesis: "A Body natural and a Body politic together indivisible; and (that) these two Bodies are incorporated in one Person, and make one Body and not divers, that is the Body corporate in the Body natural and the Body natural in the Body corporate."[81] Both figures of the King coalesce into the "Bank of England's founding as a public-private merger," which "worked, and endured, in large part because of the way in which it separated out responsibility for the dual heads-tails aspects of money. The state (heads) supplied the official government stamp that specified value. The private sector (tails) put up the store of precious metal that backed currency."[82] The trust that allows the public and private debt to merge in a single stream of sign-currency is founded in this very continuity of the Central Bank: the debt starts to pay (interest) at the same time it becomes irredeemable, infinite, and immanent. The issue we raise in this context is how the existing trust and bank could engender another trust and another common—not a central bank anymore, but a *common* one.

At the origin of the Bank of England, the dilemma consisted in the state and private dimensions of the Central Bank's role. For Massimo Amato, this dilemma was solved when it became clear that "the more Republican was the King (that is, the more he cared for the Commonwealth), the more he could gather *credit*." In other words, he was more trustworthy.[83] Consequently, paper money becomes a means of payment and an eminently net asset in England: State creditors finance public debtors more comfortably when they believe they can freely leave the credit relationship if they wish to. This was made possible by the development of a trading market for state bonds, the secondary and derivative market that soon became more important than the primary one, which deals with issuing new securities.[84] Lo and behold, the miracle of the faith in currency, the "miracle of capitalism," as Braudel defined it: "The State does not reimburse the creditors (who) regain their money at will."[85] As Bernard Schmitt points out, this is due to the fact that "currency operates in two directions simultaneously: as metal, it exerts its trading power in favor of the debtor; as *fiat* money, it holds the same power at the service of the depositor. The same money is effective twice, because the actor and his double act at the same time."[86]

Credit-money is organized around the intensification of time, with the secondary circulation of promises as a flash flood of overlaps, endorsements, and derivative launches, a money charge which spills over the social terrain, forming more or less turbulent affluences, whose driving force is constantly collected through the mills of interest, taxes, and settlements.[87] Currency stability lies in the ebb and flow, like a tide which undoes the shapes it draws in the sand through successive creative destructions, like the systoles and diastoles of a satanic lung. Concerning this process, one might ask whether the acceleration of time that allows the dematerialization of currency could, conversely, induce the deceleration of the looting of nature, of the primary and state economy of plunder. The problem is that both times are always together, dualism is constitutive of the process, just as culture and nature are invariably mingled in a single metabolism, in the constant folding and unfolding of heterogeneous processes. The temptation to solve this dilemma by acceleration (or deceleration) is completely ineffective. This is so because the infinite debt economy asymmetrically reproduces the constitutive dualism in money as debt, in its systemic dimension, and is never compelled into payment (first and foremost, the banks are the ones rescued), but it is always the other way around when it comes to populations and individuals, against whom the debts are foreclosed during acute periods of crisis. In capitalism, sign-money not only extends the subjective debt into infinity, but also creates an institution that is an endless creditor, owner of the whole deterritorialization flow that runs through society.

Researching how currency works—its nature, its mutations, and the vicious alliance it establishes with capital—allows us to understand the different strategies marshalled by the crisis theory of capitalism.[88] Most classical, neoclassical, or Keynesian theories focus on a set of practical knowledges of the art of governing aimed at rebalancing an intrinsically unstable system. This is true even if it requires the recognition of antagonistic subjectivities, as in the case of the Keynesian period that followed the crises of 1929 and World War II, when the end of capitalism was a palpable horizon in the face of the crash of the stock markets and the red menace. When it comes to Schumpeter, Hayek and, above all, the anarcho-liberalism of the Chicago School, the crisis is seen as the driving force of capitalism, so the point in question is not so much the crisis of capitalist economy, but the crisis economy, the crisis *as* economy. In the *Grundrisse*, Marx had already developed an eschatological theory of the crisis, a key moment in which the process of capital falters at the edge of the abyss, experiencing vertigo for sensing that its most creative and destitutive tendencies are unraveling. There is, therefore, a qualitative discrepancy in the positive way with which the neoliberals and non-economistic Marxists picture the crisis. Both envision a desert which is prospective solitude. The difference consists in that the former reestablish a limit for the crisis which is one step behind the absolute limit of the dissolution of capital. The latter, in turn, advocate overcoming the absolute limit through the revolution, when the sorcerer's apprentice ends up consumed in the flames he believed he could manage. This is what Deleuze and Guattari mean by "going further," overcoming the limit that implies a change in *agencement*, a new experience of time, time out of joint.[89] This is thus much different than yet another formula for rebalancing. After all, walking involves controlling imbalance, for each step is actually nothing more than a danced fall, not least because balance is impossible, as the 20th-century steel chimeras have already demonstrated, with Stalin or Mao. The attempt to even the circulation, to force the reintroduction of the theory of value so that all accounts are settled and every sum adds up, could have no other result than a totalitarian rationalization of capitalism. In fact, the socialist nightmare was born from the sleepless reason of the very political economy, when it reached its claim to maximum rationality and equivalence. The desire for order, regulation, and state, even when espoused by the Left, is nothing more than reactive nihilism that, once transposed to the political struggle, will sow authoritarian and bureaucratic reterritorializations.

In his classic book, Marc Bloch recalled that the great secret of capitalism was "to delay payments or reimbursements, and perpetually overlap the delays from ones over the others," "a system which would die if all accounts were simultaneously settled."[90] This is paradoxical, as Bloch himself points out, because the system of credit and trust, despite all previous attempts, only established itself in the 18th century, precisely when the "Lumiéres declared that all great figures of trust and faith were overcome."[91] Economists say that when trust rules, everything remains at the level of the Aristotelian *potentia*, without ever becoming act. The system is sustained by repeatedly postponing the simultaneous settlement of all accounts. Thus a dogmatic empire is born, claim the monetary economists. Deleuze and Guattari would speak in terms of an axiomatic: a monetary representation so reliable as to no longer require any confirmation or demonstration effect. At this extreme moment of capitalist subjectivation, monetary trust becomes a belief, says Amato.[92] But the inverse relationship, which is synchronic, is also true: it is belief that makes monetary trust unbreakable. However, none of that takes us very far, for we do not know the mechanisms involved in the development of trust. Sometimes, during the 20th century, this same belief was directed to the sovereign body of the maximum leader (Stalin, Hitler) instead of being focused on the money dynamics. We know the "temple" of this dynamics is the Central Bank. The Central Banks issue debts on themselves without hope of payment: it is the very debt that becomes payment. That is to say, the aggregate balance is never required, nor could it be, but the behavior of the whole is assessed at every moment. The infinite debt of monetary flows will never be collected, but individuals and populations will always be subject to it, even when it comes to weak governments, as in the case of the Vulture Funds who bought Argentinian debt and then managed to have a favorable decision of a North American judge. At the global level, there is no need for payment, ever, while it is impossible for the individual to indefinitely roll a past due debt.[93] Circulation is a vast cross-network of signs secured by other signs. We pay off our individual debt with inexhaustible global debts. Pure credit is a huge and endless path. However, at the individual level, the flow must generate a backflow (B. Schmitt); it must be redirected through the dark satanic mills, so the issue of the end violently imposes itself.

Let us go back to Rachline's scheme. One must see the world not as a trading system, but as a set of infinite and unsolvable debts that circulate without hope of extinction. An irredeemable purgatory. The traditional sequence of political economy adopts the following steps: production-surplus-trade-currency. According to the schizophrenic theory of money with which we dealt so far, the sequence would be: monetary flow-capture of wealth signs-production. Money is the starting point and not the point of arrival, like the electricity which puts the household appliances to work. Thus, economic activity is defined by the set of techniques for capturing flows used by different operators: people, states, companies, and banks, each one with its apparatus of capture and devices to temporalize debt. Its horizon does not consist of equilibrium, or even the neoliberal rhetoric of balanced accounts; it rather amounts to the opening to the becoming that currency continually renews: not zero, but epsilon; a surplus, even if infinitesimal. Currency is information and communication, a rupture in relation to the mere materiality of the coin: a pure blast with no land or god, permeated with the desire that shapes subjectivities, which unfold into different paths of actualization without ever being reduced to them, persevering within the potency regimen.[94] Regarding this, we must disagree with Orrell and Chlupaty when they define currency as quantity. A debt-claim from a bank instantly becomes currency and vice versa; money is a debt managed by the banks within economy as a whole. Currency is a credit/debt and, reversely, a debt/credit. As Bernard Schmitt would say, currency is a *non-commodity*. Sign-currency is not a sign of anything—it is pure desire. What lies behind currency? Currency itself:[95] "*Without extrinsic support, credit becomes currency*."[96] That is to say, when it comes to credit-money, difference is a condition for repetition.

As we have said before, public debt now bears interest and is *never extinct*: where liquidity is charged with trust, liquidation never happens, keeping an open future. J. K. Galbraith wrote, "Even though people desperately desire the money they put in the bank, they stop desiring it when they are sure they can get it back."[97] As Jim O'Neill, former head of global economics at Goldman Sachs, put it, the power of currency consists of a tautology: "Liquidity exists until it exists no more."[98] Suzanne de Brunhoff speaks of a structuring "concealment," while Deleuze and Guattari refer to a "cosmic trick."[99] Currency as pure sign is free from limits, but this being without end leads it to the brink of senselessness. If, for some reason, this senselessness clearly manifests itself, this is when the brink becomes the abyss where the whole monetized economy may flow to. When liquidity turns into an object of doubts and uncertainties, it then comes under pressure and, for that reason alone, puts the capitalist system on the verge of collapse. According to a passage which Keynes assigns to Lenin, "the best way to destroy the capitalist system [is] to debauch the currency."[100] This relationship between trust and liquidity can be either a *virtuous spiral* or a *meaningless tautology*. We can then say, differently than André Orléan, that the multitude is absent in both cases. When singularities cooperate among each other, trust engenders trust, forming a virtuous cycle. When singularities are reduced to fragments, trust melts into a weak tautology. In the real world, however, this is all combined, and the multitude of singularities is the crowd of fragments and the crowd of fragments is the multitude. This is where the two dimensions of time, the synchronic and the diachronic one, meet and go together, but in different manners: on the one hand, acceleration as a vicious circle of sameness, without value creation; on the other, intensive acceleration, "going further," the time of becoming related to the creation of other values: "*agencement*." There are no pure forms, no pre-constituted subjects or inescapable tendencies which we could call "multitude"; there is just a constitutive tension that conditions a possible tendency, a virtuality on which everything depends and which establishes the grounds for conflict and invention. There is only "*agencement*" and not assemblage: *making multitudes*, escaping, becoming, all in the interior of monetary/desiring flows.

For now, we may thus conclude we are dealing with the paradox of the ends, where money is an endless flow at the same it imposes an end to the resolution of the crises. The challenge is *opening* this notion of end: on the one hand, the end of the acceleration as fragmentation of social relations; on the other, the end as ethical (democratic) intensification of social relations. If the means replace the ends and no longer meet them, at the same time, ends and means coincide. Without solving this challenge, the endless acceleration will always manifest as the loss of meaning, which leads to the deadlocks and tautologies regarding the backflow and the crisis of trust. This involves our ability to unfold the dispute between the two dimensions of money (and trust) and the two velocities of credit-money. And it must be done within the cleavage that disconnects the systemic debt (at the macro level), which will never be acquitted, from the one that reduces subjectivation to subjection (at the micro level). In other words, trust is a sensible projection of the becoming into an infinite debt in the face of the belief in a senseless future. It is no coincidence that Rachline advocates the coordination of global interdependency—and thus a utopian World Central Bank as the *locus* for the conciliation of flows—while opposed to the independence of such bank, for otherwise it would mean it would work as a second nature, even more arbitrary than the first one. Would it be possible to reclaim banks and finance beyond capitalist capture? Would it be possible to unblock the monetary flows of the alliance with capital? How would the new currency work in logical terms? And, more importantly, what would the historical and political conditions for its creation be?

NOTES

1. *Debt*, p. 8.
2. Tim Marshall, *Prisoners of Geography*.
3. We know Parsons used to ask himself, "Why, then, did the breakthrough to modernization not occur in *any* of the 'Oriental' advanced intermediate civilizations?" Talcot Parsons, *Societies: Evolutionnary and Comparative Perspective* (New Jersey: Prentice-Hall Incorporated, 1966), 44.
4. Jack Goody, *The East in the West*. Goody rests on the works of R. de Roover. See Roover, "Aux origines d'une technique intellectuelle. La formation et l'expansion de la comptabilité à partie double," *Annales d'histoire économique et sociale* 9, no. 45 (May 31, 1931): 270–98, http://www.jstor.org/stable/27574484. See also John Kenneth Galbraith, *La moneta: da dove viene e dove va*, trans. Capriolo (Milan: Mondadori, 1976), 29. The usual reference is Frederic C. Lane, "Venetian Bankers, 1496–1533: A study of the early Stages of Deposit Banking," *The Journal of Political Economy* 45, no. 2 (April 1937): 187–206.
5. David Orrell and Roman Chlupaty, *The Evolution of Money* (New York: Columbia University Press, New York: 2016), 58.
6. Ibid., 9. The authors also write that "(monetary transactions) involve an exact amount—you can put a number on them."
7. Goody, *L'Orient en Occident*, 76–79.
8. Orrell and Chlupaty, *The Evolution of Money*, 57. This was the Franciscan friar Luca Pacioli, who wrote *Summa Arithmetica* in 1494.
9. Jack Goody, *La raison graphique*, 86–87.
10. Orrell and Chlupaty, *The Evolution of Money*, 58.
11. Ernest Kantorowicz, *The King's Two Bodies: Studies in Mediaevel Political Theology* (New Jersey: Princeton University Press, 1997 [1957]).
12. Regarding money as flow, it is worth noting that one of the etymologies of the word money (DaMim) in Hebrew is DaM, blood. *Cf.* Emeric Deutsch, *L'argent: Données et débats. Actes du XXVIIIe colloque des intellectuels juifs de langue française* (Paris, Denoël: 1989), 82.

13. It is relevant to mention here the debate over forms of ownership and control of *corporations*, which does not fit into the scope of our analysis, but represents a very important development. See, for instance, the proposal in terms of passive and active property in Gardiner Means and Adolf Augustus Berle, *The Modern Corporation & Private Property* (New Brunswick: Transactions Publishers, 1968).

14. "To code desire—and the fear, the anguish of decoded flows—is the business of the socius." Gilles Deleuze and Félix Guattari, *Anti-Oedipus: Capitalism and Schizophrenia* (Minneapolis: University of Minnesota Press, 1983 [1972]), 139. Still according to Deleuze and Guattari, the "despotic machine holds the following in common with the primitive machine, it confirms the latter in this respect: the dread of decoded flows—flows of production, but also mercantile flows *(flux marchands)* of exchange and commerce that might escape the State monopoly, with its tight restrictions and its plugging of flows." *Anti-Oedipus*, 196.

15. Bernard Schmitt, *Monnaie, Salaires et profits* (Albeuve: Castela, 1975 [1966]), 157.

16. John Maynard Keynes, "Economic possibilities for our grandchildren (1930)," in *The Collected Writings of John Maynard Keynes. Volume IX: Essays in Persuasion* (Basingstoke: Palgrave Macmillan, 2010), 324.

17. John Kenneth Galbraith, *La moneta: da dove viene e dove va*, trans. Capriolo (Milan: Mondadori, 1976), 24. Originally published in English: *Money: Whence It Came, Where It Went* (1975).

18. Ibid., 16.

19. André Amar, "Essai psychanlitique sur l'argent," *Revue française de psychanalyse* 20, no. 3 (1956): 332–44, http://bsf.spp.asso.fr/index.php?lvl=notice_display&id=13586.

20. For it is not enough to celebrate the decoding flows in some sort of euphoria for the dissolution of social forms and the opening of a broad horizon of transformations without considering the *confrontational* and *violent* aspects that are in constant modification. As we have examined on chapter 2, the whole discussion around uneven development, dependency, deterioration of the terms of trade, or asymmetrical interdependency show how modernity sublates archaic aspects, the public sublates the private, new markets and freedoms sublate violence and racism. In this sense, Deleuze and Guattari say, "The action of decoded flows is not enough, however, to cause the new break to traverse and transform the socius—not enough, that is, to induce the birth of capitalism. Decoded flows strike the despotic State with latency; they submerge the tyrant but they also cause him to return in unexpected forms; they democratize him, oligarchize him, segmentalize him, monarchize him, and always internalize and spiritualize him." *Anti-Oedipus*, 223.

21. "[T]he structural exchangist conception tends to postulate a kind of primary equilibrium of prices, a primary equivalence or equality in the underlying principles, which allows it to explain that the inequalities are necessarily introduced in the consequences." Deleuze and Guattari, *Anti-Oedipus*, 187.

22. See chapters 3 and 4.

23. Étienne Balazs, *Chinese Civilization and Bureaucracy: Variations on a Theme* (New Haven: Yale University Press, 1964), 11.

24. *Anti-Oedipus*, 197.

25. Karl Marx, "Reply to Milkhailovski" (November 1877), in Karl Marx and Friedrich Engels, *Basic Writings on Politics and Philosophy* (Garden City, N.Y.: Doubleday, 1959), 441. *Apud* Deleuze and Guattari, *Anti-Oedipus*, 223.

26. In turn, this "Oriental" tyrannical continuity conditions a scheme of struggles based more on rebellion than on revolution—that is, in the constant confrontation against an almighty state (anarchist typology) to the detriment of the dialectical seizure of state power to transform and occupy it as a necessary step for a classless society. "This is why the order of latency in the African, Chinese, Egyptian, and other empires was that of rebellions and constant secessions, and not that of revolution. Here again, death will have to be felt from within, but it will have to come from without." Deleuze and Guattari, *Anti-Oedipus*, 219. Still according to Deleuze and Guattari, "Marx's observations on the despotic formations of Asia have been confirmed by the African analyses of Max Gluckman, *Custom and Conflict in Africa* (Glencoe, 111: Free Press, 1959): at the same time immutability of form and constant rebellion. The idea of a 'transformation' of the State indeed seems to be a Western one. And that other idea, the 'destruction' of the

State, belongs much more to the East and to the conditions of a nomad war machine." Deleuze and Guattari, *A Thousand Plateaus: Capitalism and Schizophrenia*, trans. Brian Massumi (London/Minneapolis: Minnesota Press, 1987 [1980]), 558.

27. Étienne Balas, *La bureaucratie celeste* (Paris, Gallimard, 1968).

28. Here is another example: "Between 1661 and 1663, the Qing reimposed the ban on private sea traffic and pursued a scorched-earth policy that transformed China's southeast coast from a crucial link connecting the Chinese and the world markets into a no-man's-land that kept the two apart." Giovanni Arrighi, *Adam Smith in Beijing: Lineages of the Twenty-first Century* (London-New York: Verso, 2007), 326.

29. In this sense, as a theory of analysis of capitalism, Marxism either is a monetary theory of value or it is not Marxism. See John Milios, Dimitri Dimoulis, and George Economakis, *Karl Marx and the Classics: An Essay on Value, Crises and the Capitalist Mode of Production* (Burlington: Ashgate Publishing, 2002), vii, 10, 28, 127.

30. "The decoding of flows and the deterritorialization of the socius thus constitute the most characteristic and the most important tendency of capitalism." Deleuze and Guattari, *Anti-Oedipus*, 34.

31. According to Deleuze and Guattari, Schmitt carried the reflection on the schizoid nature of this second flow, inconvertible in wealth as such, to the end. As the authors put it, "'nonpossession and nonwealth,' in the words of Bernard Schmitt, who among modern economists has for us the incomparable advantage of offering a delirious interpretation of an unequivocally delirious economic system (at least he goes all the way). In short, a truly unconscious libido, a disinterested love: this machine is fantastic." Deleuze and Guattari, *Anti-Oedipus*, 374.

32. It is the very mutation of currency that occasions the emergence of a capital which is not merely external in relation to economy, but infiltrates the monetary process and, carried by it, reaches the desiring flow. "Before the capitalist machine, merchant or financial capital is merely in a relationship of alliance with noncapitalist production; it enters into the new alliance that characterizes precapitalist States—whence the alliance of the merchant and banking bourgeoisie with feudalism. In brief, the capitalist machine begins when capital ceases to be a capital of alliance to become a filiative capital." Deleuze and Guattari, *Anti-Oedipus*, 227.

33. "Hence one is correct in speaking of a profound *dissimulation* of the dualism of these two forms of money, payment and financing—the two aspects of banking practice." Deleuze and Guattari, *Anti-Oedipus*, 229.

34. For a slightly different version of this typical equation regarding the Bergsonian ontology of variation, see *Anti-Oedipus*, 226–31.

35. In *Anti-Oedipus*, Deleuze and Guattari call it "machinic surplus value" (msv), distinguishing it from the human surplus value. The msv is the affirmation of the direct productive power of the machination between human and machine, between nature and culture. For the authors, capitalism rests on the generalized machinism of the productive unconscious, so the human surplus value is a pre-capitalist element. Although the notion of msv contradicts the Marxist principle according to which the surplus value stems from the exploitation of human labor, it is not possible to derive surplus value from constant capital (in the machines). This concept is derived by Deleuze and Guattari from a delirious reading unraveled from Marx himself—more specifically, from the passage known as "Fragment on machines" in the *Grundrisse*. See also Antonio Negri, *Marx Beyond Marx: Lessons on the Grundrisse* (Brooklyn, New York: Autonomedia, 1991).

36. See Milios, Dimoulis, and Economakis, *Karl Marx and the Classics*.

37. Marx, *Grundrisse*.

38. François Rachline, *Que l'argent soit: Capitalisme et alchimie de l'avenir* (Paris: Calmann-Lévy, 1993), 36. There are six different cycles of capital accumulation which correspond to the global economic emergence of six cities: Florence, Genoa, Venice, Amsterdam, Paris, and London. In this succession, we notice the passage from city-states to capital cities of states which will play a crucial role in the colonization of the world.

39. Christian Kerslake, "Marxism and money in Deleuze and Guattari's Capitalism and Schizophrenia: On the Conflict between the Theories of Suzanne de Brunhoff and Bernard Schmitt," *Pahrresia*, no. 22 (2015): 38–78, http://www.parrhesiajournal.org/parrhesia22/parrhesia22_kerslake.pdf.

40. Pierre Klossowski, *Living currency* (London: Bloomsbury Academic, 2017 [1970]).

41. "For it is certain that the regime of decoding does not signify the absence of organization, but rather the most somber organization, the harshest compatibility, with the axiomatic replacing the codes and incorporating them, always *a contrario* Deleuze and Guattari, *Anti-Oedipus*, 153. See also Gilles Deleuze, "Postscript on the Societies of Control," *October* 59 (Winter 1992 [1990]): 3–7, http://www.jstor.org/stable/778828.

42. Michel Foucault, *Le gouvernement de soi et des autres* (Paris: Gallimard, 2008).

43. Maurizio Lazzarato, *The Making of the Indebted Man: An Essay on the Neoliberal Condition* (Los Angeles: Semiotext(e)/Smart Art, 2012).

44. Deleuze and Guattari, *Anti-Oedipus*, 239, 240. Since mid-2016, the *Obsolete Capitalism* group has edited a series of articles, interviews, and booklets around this passage and Nietzsche's original accelerationist fragment. Available at http://obsoletecapitalism.blogspot.com.br/.

45. "One of Keynes's contributions was the reintroduction of desire into the problem of money; it is this that must be subjected to the requirements of Marxist analysis. That is why it is unfortunate that Marxist economists too often dwell on considerations concerning the mode of production, and on the theory of money as the general equivalent as found in the first section of *Capital,* without attaching enough importance to banking practice, to financial operations, and to the specific circulation of credit-money—which would be the meaning of a return to Marx, to the Marxist theory of money." Deleuze and Guattari, *Anti-Oedipus*, 230. However, as we shall discuss in the next subchapter, the marginalists were the ones who introduced, long before Keynes, the desire in the economy.

46. John Maynard Keynes, "Auri Sacra Fames (1930)," in *The Collected Writings of John Maynard Keynes. Volume IX: Essays in Persuasion* (Basingstoke: Palgrave Macmillan, 2010), 163. Emphasis added.

47. Keynes, "The End of the Gold Standard (27 September 1931)," in *Collecter Writings*, 245–50.

48. "Money is coined liberty, and so it is ten times dearer to a man who is deprived of freedom. If money is jingling in his pocket, he is half consoled, even though he cannot spend it." Fyodor Dostoyevsky, *The House of the Dead* (Mineola, New York: Courier Corporation, 2014), 14.

49. Keynes, *The General Theory of Employment, Interest, and Money* (Macmillan and Co., 1964).

50. Orrell and Chlupaty, *The Evolution of Money*, 176.

51. Deleuze and Guattari, *Anti-Oedipus*, 218.

52. The first accelerationist manifest was the Manifesto of the Communist Party (Marx and Engels, 1848).

53. Once again, Deleuze and Guattari stress the profound ambiguity of the death instinct when it directly concerns the subjectivity production in capitalism: "The dissolutions are defined by a simple decoding of flows, and they are always compensated by residual forces or transformations of the State. Death is felt rising from within and desire itself becomes the death instinct, latency, but it also passes over into these flows that carry the seeds of a new life. Decoded flows—but who will give a name to his new desire?" Deleuze and Guattari, *Anti-Oedipus*, 223.

54. Deleuze and Guattari, *Anti-Oedipus*, 10.

55. "We shall speak of an *absolute limit* every time the schizo-flows pass through the wall, scramble all the codes, and deterritorialize the socius: the body without organs is the deterritorialized socius, the wilderness where the decoded flows run free, the end of the world, the apocalypse." Deleuze and Guattari, *Anti-Oedipus*, 170. From this results the main organizational proposition of the book: "And if materialist psychiatry may be defined as the psychiatry that introduces the concept of production into consideration of the problem of desire, it cannot avoid posing in eschatological terms the problem of the ultimate relationship between the analytic machine, the revolutionary machine, and desiring-machines." *Anti-Oedipus*, 35.

56. Schmitt, *Monnaie, Salaires et profits*, 157. Emphasis added.

57. "Capitalism is in fact born of the encounter of two sorts of flows: the decoded flows of production in the form of money-capital, and the decoded flows of labor in the form of the 'free worker.'" Deleuze and Guattari, *Anti-Oedipus*, 33.

58. The global free-market technology of the British Empire operated along with the technologies of its fluidized war machine, including in the strategical field with the *fleet in being*, when a naval force extends a controlling influence without ever leaving port (which paradoxically involves being defeated if the circumstances force it so). "The sea as a smooth space is a specific problem of the war machine. As Virilio shows, it is at sea that the problem of the *fleet in being* is posed, in other words, the task of occupying an open space with a vertical movement that can rise up at any point." Deleuze and Guattari, *A Thousand Plateaus*, 383. See also Guillaume Sibertin-Blanc, *State and Politics: Deleuze and Guattari on Marx*, trans. Ames Hodges (Los Angeles: Semiotext(e)/Smart Art, 20126), 112.

59. Carl Schmitt, *Land and Sea*, trans. Simona Draghici (Washington, D.C.: Plutarch Press, 1997), 50–51.

60. Giovanni Arrighi, Giovanni Arrighi, *The Long Twentieth Century: Money, Power and the Origins of our Time* (New York: Verso, 1994).

61. Rachline, *Que l'argent soit*, 70.

62. Clément Rosset, *L'anti-nature*, 5th ed. (Paris: PUF, 1973), 18.

63. Rosset, *L'anti-nature*, 28.

64. Orrell and Chlupaty, *The Evolution of Money*, 106.

65. "Currency is created based on the projection of a credit-debt." Schmitt, *Monnaie*, 189. More than projection, we understand currency as constitution, forming a constitutive assemblage with desire.

66. Schmitt, *Monnaie*, 188.

67. The dramatic aspect of this challenge is that, today, possession or lack of money are the most apparent facets of inequality and exploitation throughout the world, which tends to refer the analysis to the moral plane, one way or another (money saves *and* corrupts).

68. In other terms, this is Antonio Negri's thesis of real subsumption as topical issue of modern capitalism. Deleuze and Guattari would speak in terms of machinic capitalism and Foucault, biopolitical or securitarian society. See Matteo Pasquinelli, ed., *Alleys of Your Mind: Augmented Intelligence and Its Traumas* (Lüneburg: Meson Press, 2015). Available at http://matteopasquinelli.com/?p=2872.

69. Orrell and Chlupaty, *The Evolution of Money*, 112.

70. Georg Simmel, *Philosophie des Geldes*, trans. Sabine Cornille eand Philippe Ivernel, *Philosophie de l'argent* (Paris: PUF, 1987 [1900]).

71. Simmel, *Philosophie des Geldes*, 195.

72. André Orléan, *The Empire of Value: A New Foundation for Economics*, trans. M. B. DeBevoise (Cambridge: The MIT Press, 2014), 144. Further on, we shall return to Orleán's proposal for a currency approach in terms of mimetic polarization.

73. Simmel, *Philosophie des Geldes*, 197.

74. Peter Linebaugh and Marcus Rediker, *The Many-headed Hydra: Sailors, Slaves, Commoners, and the Hidden History of the Revolutionary Atlantic* (Boston: Beacon Press, 2000).

75. Rachline, *Que l'argent soit*.

76. Antonio Negri, *Potere Costituente. Le alternative della modernità* (Milan: Sugar&co: 1992).

77. "The very notion of a monetary mass can have a meaning only relative to the workings of a system of credit where the different kinds of money combine. Without such a system, one would have only a sum of means of payment that would have no access to the social nature of the general equivalent and that could serve only in local private circuits. There would be no general monetary circulation. Only in the centralized system can the different kinds of money become homogeneous and appear as the components of an articulated whole." Suzanne de Brunhoff, *L'offre de monnaie: Critique d'un concept* (Paris: Maspero, 1971), 124, apud Deleuze and Guattari, *Anti-Oedipus*, 230. For a close reading of Deleuze and Guattari's approach to this French neo-Marxist economist, see Kerslake, "Marxism and money."

78. Rachline, *Que l'argent soit*, 162. Rachline, as well as many writers on economy or history, always refer to the classic work of Ernest Kantorowicz.

79. Rachline, *Que l'argent soit*, 438.

80. Ibid., 423.

81. Ibid.

82. Orrell and Chlupaty, *The Evolution of Money*, 92.

83. Massimo Amato and Luca Fantacci, *Fine della finanza: da dove viene la crisi e come si può pensare di urscirne*, 2nd ed. (Roma: Donzelli, 2012), 213.

84. Amato and Fantacci, *Fine della finanza*, 123.

85. Fernand Braudel apud Rachline, *Que l'argent soit*, 469.

86. Schmitt, *Monnaie*, 128.

87. Reference to the "dark satanic mills" from William Blake's poem "Jerusalem," with which he designates the first factories created in the context of the development of the capitalist factory system, in the late 18th century, one of the first and main expressions of the valorization of abstract time. The expression will be taken up by the Austrian economist Karl Polanyi to describe the gears of the industrial revolution in England during the 1800s. Cf. Karl Polanyi, *The Great Transformation: The Political and Economic Origins of Our Time*, 2nd ed. (Boston, MA: Beacon Press, 2001 [1944]).

88. Antonio Negri, *Revolution Retrieved: Writings on Marx, Keynes, Capitalist Crisis, and New Social Subjects (1967–83)* (London: Red Notes, 1988).

89. "The time is out of joint. O cursed spite, / That ever I was born to set it right!" Shakespeare, *Hamlet*, act 1, scene 5.

90. Marc Bloch: *Esquisse d'une histoire monétaire de l'Europe* (Paris: Armand Colin, 1954), 77.

91. Jean-Michel Rey, *Le temps du crédit* (Paris: Desclée de Brouwer, 2002), 24.

92. Ibid., 185–94.

93. Ibid., 185.

94. Ibid., 196.

95. Ibid., 199.

96. Ibid., 110.

97. Galbraith, *La moneta*, 27.

98. Apud Amato e Fantacci, *Fine della finanza*, 33.

99. Marx, in turn, ascribed the "metaphysical subtleties and theological gimmiks" to commodity. We believe this is a much more appopriate definition for the nature of currency.

100. Keynes, John Maynard. *The Collected Writings of John Maynard Keynes. Volume 1: The Economic Consequences of the Peace* (London: Macmillan, 1971 [1919]).

Chapter Six

Anthropological Radicalism of Contemporary Capitalism

MONEY BEYOND VALUE

Currency inhabits the tension between the material and immaterial, support and sign, synchronicity of exchanges and diachronicity of debts, wealth representation and pure deterritorialized flow. Its relation with sovereign power has always been ambivalent. If, on the one hand, the states introduced currency within a vertical circuit and there they kept it with all their might, on the other, there is a throbbing deterritorialization tendency that goes beyond the codes of state bureaucracy. In the interstate wars, in the plundering of populations, in the colonial looting, currency is appropriated by the State as the direct spoils of its subjugation power. The stored metal thus represents an amount directly proportional to the monopoly on the violence of appropriation exerted by state power. By breaking with the point of comparison which links it to the material standard and allows it to be measured by the sovereign power, the mutation of the currency winds up eroding this power, draining its capacity to mobilize and rule the *socius*, until it is completely swallowed by the deterritorialization of social productivity. Hence the constant vigilance, the regulation on the use of money; the point is to prevent it from indirectly acting as a solvent of the secure conditions to which the state applies its nets and measures of royal science in an obsessive reaffirmation of its sovereign power. No wonder it imprints its paranoid face in each circulating coin. The social-evolutionist approach generally explains the emergence of stocks as a sign of development of the agricultural production forces, which would have enabled primitive economies to fulfill their subsistence need, additionally allowing the accumulation for the future. According to this teleology, there is a transition from a primitive communism of collective property toward the first states, which built up the political power and organized around the conquest and safekeeping of stocks. However, the stock innovation did not occur just to accumulate surplus goods that would have no use in the present. The stock is not a positivity that would supply the lack of surplus; it actually precedes this absence, introducing and organizing it—a logic by which the state-form is already operating. In the philosophical archeology of the state developed by Deleuze and Guattari in *A Thousand Plateaus*,[1] the stock-form is a qualitative invention, a structural condition that changes the nature of the circulation productive regime. Paradoxically, the state precedes and at the same time results from this innovation, as if it presupposed its own ideality, like the self-movement of the idea toward the concept, in Hegel.[2] According to the authors, the stock was not engendered by technical evolution and then became the object of disputes, leading to the coagulation of political power. Instead, a separate political power emerged by provoking the cleavage of the stock-form between activity and surplus, or land and proprietary income. The state does not arise from a progressive transformation of the production mode; it actually occurs at once, by changing the production assemblage. As Deleuze and Guattari put it, "It is not the State that presupposes a mode of production; quite the opposite, it is the State that turns production into a 'mode.' The last reasons for presuming a progressive development are invalidated."[3] The state behaves as an apparatus of capture in three spheres, since the stock innovation introduces the possibility of a threefold quantitative comparison: *1*) among lands (through the less productive one), *2*) among activities (through surplus work), and *3*) among measures (through the equivalence principle, which justifies taxation). Once the measurement standard is established, the equivalence principle emanates from the very principle of state sovereignty, so that the state and its original appropriation of lands, activities, and money—something Marx theorized in *Capital* under the concept of "primitive accumulation," are already embedded into each measurement or comparison.

Building on Pierre Clastres's ethnographic anarchism,[4] Deleuze and Guattari argue that primitive societies deliberately ward off the formation of a distinct and superior political power. The effort of inhibiting the state, this terrible power which menaces the social body, works through a set of anti-political strategies and social counter-powers, whose goal is to permanently disable the formation of stocks and the consequent indexation of productivity by the royal measure and its equivalence system. In societies without state, the social body must remain sufficiently ritualized and codified in order to prevent said societies from crossing the qualitative threshold that implies the arrival of the state. When it occurs, the state arrives at once, overcoding the flows. This *anti-production* mechanism, this cosmological anti-politics, is actually a positive element, for it consists of complex practices aimed at dissolving emerging hierarchies, avoiding the detachment of a distinct power, and deactivating processes of concentration of power. Among some examples of this constant practice of warding off the state, the authors follow Clastres's footsteps and mention primitive wars, anthropophagy, and rituals of destruction, gift-giving, or unrestrained consumption (the potlach, for example). Therefore, the stock is not the result of the development of productive forces, but a political defeat imposed on primitive societies, whose subjugation causes the state to emerge, dissolving their constitutive horizontality. The stock does not evolve from barter; on the contrary, it supposes barter once state power traverses it. That is to say, there was no economic world based on the exchange of equivalents among equals: the very equivalence results from the state imposing itself and implies inequality from the start.

In *A Thousand Plateaus*, the authors write that "exchange does not assume a preexistent stock. . . . Stockpiling begins only once exchange has lost its interest, its desirability for both parties."[5] This lack of interest within exchanges, the absence brought into desire, that is what digs the void where state power installs itself. This is when stocks begin to determine value, at the receding moment of desire, when desirability is replaced with equivalence, a common measure, when the interest in the relationship falters. This schizoanalysis of the state has nothing to do with an evolutionism of productive forces, according to which the state would be born as the result of the formation of stock amid the communal mode of production. Instead, it is about the very self-founding of the state as stock-form.[6] Under the currency theme, this means that the foundation of the state coincides with the foundation of a common measure, a term of comparison for land income, surplus work and taxation, all that is tied to the vertical circulation circuit, whose apex is the sovereign. We define the ether of this economy of exchanges ruled by the sovereign power in one word: *value*. There is no difference between value and surplus value, for the latter (appropriation, expropriation, direct pillage) precedes the former (measure, equivalence, standards). Therefore, value is not a properly capitalist production, but a state invention that will be later reconfigured by the process of capital. In every case, value carries the threatening effigy of the sovereign as well as a latent and sinister terror, and jealously guards each and every transaction. Surplus value—the original violence that enforces the law, the primitive accumulation– not only precedes value, but also joins it in all transformations it undergoes through history, in all its political and economic rearrangements. State power institutes value—which has always been an objectified value with a certain degree of violence that must be constantly reproduced—not the spontaneity of exchanges. Value implies the verticality that dissolved primitive societies as a comparative term for quantities and a common measure for heterogeneous series, overdetermined by the state from that point on. Measure only comes into force in history through the violence it carries in the womb, the terrifying face of the original state-form.

During the industrial revolution, the final period of capitalist formation in the West, classical political economists adopted the centrality of the notion of value in a neutral and objective way for the sake of establishing a scientific criterion that would work as a secondary grounds for the new science. On the one hand, they were trying to inaugurate a practical science of government beyond the legal model of contractualist political theories (Rousseau, Locke); on the other hand, they understood that the concepts adopted by English utilitarianism—as in Bentham—were too subjectivist to describe the process of wealth production. In this context, David Ricardo developed the theory of value whose main output is to determine how to objectively calculate the value resulting from the crystallization of social labor time in a given product or service. That is the first theory which can be orchestrated for distributive purposes, since it allows measuring goods, services, and activities according to the equivalence principle. This theoretical effort coincides with the political effort to maintain the state overcoding of currency, at a time when the mercantile, industrial, and financial circuits of society seemed to leak from all sides, like gas escaping under high pressure from an inadequate valve. The classical theory of value, a future favorite and rationalist trope of social-democratic and bureaucratic-socialist theories, emerges with an immediate political role: keeping money within the vertical circuit frame, the quantitative system of the common measure which allows the state to quantify the three spheres it captures: land, labor, and taxes. This curious coincidence of economic liberalism with the imperative reaffirmation of state power should not come as a surprise if we consider that the Smithian principles of *laisser faire* or *ne gouverner pas trop* are not to be confused with an *ex-post* biased reading that equals Smith to an apology for the "invisible hand," read as a mystical self-regulation of markets. There is no doubt the political economists have theorized limits and conditions for government action, above which its intervention in the economy would compromise the supposed virtuous rationality of free agents, acting in their own self-interest. Yet, by no means have they claimed the absence of boundary conditions for the balanced workings of the market, which admittedly depends on a centralized political power determined to intervene when such conditions are at risk. Among those conditions, the centrality of *value* is evident in Ricardo's case. He invokes a convenient and rational measure that allows for a criterion of distributive justice, which already implies a philosophy of the state-form: the equivalence principle, the common measure of wealth, and the overcoding of money.

Hence classical political economists understood currency essentially as a facilitator for trades, an instrument whose quantity should be calibrated to ensure the ether of economy, as if it entered the economic circuit from outside the productive cycle. Money would not be productive in itself, but an aid. Such premise is directly linked to the postulate of general economic balance (especially in Jean-Baptiste Say), in which inputs and outputs only increase due to labor, to the real production of converting natural into artificial matter. In this sense, unlike Marx, balance is the rational and orderly premise, the "natural state" in the light of reason, while imbalance is the result of irrational and anarchic charges. The postulate of equilibrium, coupled with the theory of value, extends the vigilance of state power over the potential excesses of currency, over the reasons of its unreasonable behavior, the straight and lucid measure of the placid economic scientist against human, all too human, passions and whims. Value is thus the watchdog standing next to currency. At the same time that currency was being deterritorialized and fermenting the industrial revolution, the most advanced economic theories were locking them up in the state stock. Political economy thus follows the state power in the effort to attach an absence to the heart of currency (value), a measure to boundless desire, a qualitative anchor to the heterogeneous whims of humanity, so that the mutation into sign can be prevented and money may keep on being treated as a facilitator, an ether or, worst case scenario, a commodity. Hence the perception that economy would have a demand and a supply *for* currency, and the political power, in its macroeconomic mission, would be responsible for balancing and fine-tuning it: "an economy in need of constant monetization."[7]

According to André Orleán, we live under an "empire of value."[8] From the point of view of value, exchanges would be universally valid and money in itself would have a secondary role.[9] Exchanges would be a natural operation, some sort of fair commutation based on the nature of things. Nobody loses: the final result is zero, tit for tat, *suum cuique tribuere*. The balanced exchange logic is the peace of God and Reason. That way, the equivalence principle becomes the principle of rationality and intelligibility to solve the mystery of monetary tautology. This kind of reasoning, of Aristotelian-Thomist background, condemns the monetary arrangements beyond simple exchanges as artificial, deceptive, and transgressive of the due measure. Beyond the supposed equivalence of simple trade, there would be nothing more than illegitimate gain, parasitic speculation, excessive usury, the hubris of financial horse-trades, and sheer theft. This is the long history of "material populism" and the back-to-basics cries, both on the Left and Right of the political and ideological spectrum—going from the socialist Left to the Tea Party and Donald Trump, from the Latin American developmentalists to the 70% certified Maoist capitalism of Deng Xiaoping—which will end up in the coalition of moral accusations toward finance—this speculative casino operated by degenerated yuppies—as a distortion of the real economy, the "productive sector," and capitalism well-grounded on work and headed by good employers.[10] Thus work would also have a due measure, a fair wage that would correspond to the amount of effort, time and qualification crystallized in it. Hence the strange sympathy the socialists have for a "capitalism without capitalists," meaning the problem was not the (human and machinic) exploitation, but the excessive profit, the disruption of the equivalence principle, the disrespect for the rationality of value. No wonder the maximum accomplishment of the Ricardian theory of value did not occur in the West, but in the Stalinist USSR, with its infernal quotas, tables, and unappeasable bureaucracies. Neither was the American self-made man the most perfect subjective accomplishment of puritanical and ascetic ethics, but the Soviet Stakhanovism, the marble champion carved for the "real economy" and its productivist premise. In his notebooks on Marx, Negri says that the theory of value is used as a club against us.[11] It is the architrave of the objectivist Marxism, its instrument for order purposes. It does not matter whether we are dealing with economy-focused Marxists, classical or neoclassical economists, whenever the theory of value is enthroned as the foundation for all measures, it works against the deterritorialization of money (its de-foundation) on behalf of bureaucratic instances and hierarchical codifications of the state.

The approach to value as a measurable objectification of labor does not allow for the monetary dualism of the lines. Hence we must turn the classic approach upside down: it is not currency and finance that mystify value and real economy; on the contrary, value and real economy are the ones that conceal the real workings of money. Keeping within the theory of value framework does not contribute to the understanding of monetary institutions as capital policy, including their state dimension, since, in capitalism, the states are readapted for the purpose of sustaining the monetary and desire flows. In the first paragraph of the introduction to his book on currency, Bernard Schmitt says: "Although the 'Prince of the economists,' David Ricardo, has started by studying currency, the foundation of classical economic theory lies in *value*, not in currency." Since the 1870s, however, "the so-called 'neoclassical' economists refused the Ricardian theory of value for being . . . absolutist and mythical, when every actual value is indeed relative." The author is referring to the emergence of the marginalist revolution, with the publication of works of economists such as W. S. Jevons, Auguste and León Walras, and Carl Menger. As he puts it, "despite this revolution of thought, economic science still follows the same method, i.e., that the notion of value *must* precede everything."[12] Currency is relegated to a secondary role regarding value. At the end, even the marginalist revolution of the 1870s did not entirely escape from repeating a hypostasis in order to substantiate the economic science: labor is then replaced with marginal utility. They introduce another substance as something that exists a priori, a safe ground to erect the theoretical building. "First of all, reality; then, the monetary aspects. . . . They assemble a world they call 'real' and attach the monetary component to it."[13] Reality is considered authentic and the monetary (the artificial) aspects are reduced to the fictitious level. Neoclassical economics does not leave the classical view completely behind, for it still maintains, even if residually, the separation between a real world and a monetary one, trying to build then a bridge between them.

Orleán proposes an ultimate critique of the substantialist approaches to the notion of value, whether classical (labor-value) or neoclassical (utility value). He concentrates his analysis on neoclassical and marginalist theories, given their hegemonic character within the financial system and amid academic studies in economics. Moreover, we believe it is important to understand those theories for criticizing not just them, but our own formulations, since the neoclassical economists were able to delve much deeper into the workings of capitalism than the theories based on labor-value. Thus, Orleán unravels more than the basic axioms of the neoclassical and marginalist theory; he also explores the answers that the economists gave throughout history regarding the shortcomings of the utility theory, so it will keep on working and dealing with crises. Those efforts, as the economist of the French Regulation School points out, far from solving the deadlocks of neoclassical economics, show that we are not dealing with natural or naturally rational phenomena; we are actually facing political and institutional arrangements.[14] But Oreleán's critique, however consistent it may be within the problematic field he explores, circumvents the main potential of the marginalist thought, its extraordinary intuition. Without grasping it, his critical blade loses the edge and winds up reduced to the moral plane, to the good or bad use of state control mechanisms, as it repeatedly occurs with left-wing analysis, which never tires of spinning around inadequate and normative categories, usually regarding some critical scheme against neoliberalism.[15] Thereby one misses more than the image of the marginalist thinking; one loses its touch point with the very imaginary of capitalism. To fill in this gap, it is worth revisiting Jean-Joseph Goux's research on the marginalist revolution as intuition of the great rupture that founds our present.[16]

Goux's originality lies in placing marginalist economics within an epistemic shift that involves not only economics, but also philosophy, painting, literature, and the society in general. At a major epistemic bifurcation by the end of the nineteenth century, the marginalists traveled the same river that impressionist paintings and Mallarmé's poetry did. The point was no longer painting the object, but the effect it produced.[17] As the French economist brings up, the pioneer work of Léon Walras and the article that first employed the term "impressionism," in a derogatory reference to one of Monet's paintings,[18] suggestively date back to the same year, 1874. As impressionist painters like Cezánne put aside the program of representing the thing itself on behalf of painting its sensible effects—the sensation logic within the gray zone between subjective and objective—the first neoclassical economists devised an "economic impressionism" by refusing the principles of pure reason implied in the objective approach to value and equivalence.

Goux thus explains how the marginalists were the first to perceive sign-currency as the engenderer of foundational ruins, as the desert that advances with the death of transcendences, in accordance with a critique of dogmatic metaphysics that brings them philosophically closer to Nietzsche's active nihilism as opposed to the Platonic or late Platonist tradition. In this sense, the stock exchange model does not explain the financial moment as sheer excess or a mere accessory of capitalism, but as its own dynamic and essential core, exactly where the marginalists place the turbo-engine of desire. And the workings of such desire may be verified, without any puritanical judgment, within the fashion world, in mannerisms, and in the storefronts of consumer appeal by which Walter Benjamin wandered—all of that which emerge at that same financial moment, at the end of the nineteenth century. In a sense, by the last decades of that century, capitalism was already advanced. The marginalist revolution represented a shift based on the very formation of capitalism, capturing a component that went unnoticed by classical economists. Regarding that, the marginalists could anticipate in at least a century the whole post-modernist theoretical framework. Instead of placing the dissolution of the theory of value in the 1970s—for instance, in the transition from the (Keynesian-Fordist) plan-state to the (Schumpeterean-neoliberal) crisis-state[19]—Goux refers to the fact that this argument was already present, in a very consistent and sophisticated way, in the seminal works of the economics thinking of the 1870s. Now, amidst the crises of post-Fordist regulation and the debates on cognitive capitalism and the post-industrial society, one can retrospectively ascertain the fertility of Walras's, Jevons's, and Menger's intuitions.

Regarding our problem, it is important to mention the criticism which Goux mobilized, through the recovery of marginalists and impressionists, opposing the claims against the artificial, the fictional, and the subjective on behalf of economic or economistic objectivisms, such as the morals implicit within the Ricardian theory of value. Essentially, this moral leans on a realistic and naturalistic metaphysics of needs. This concerns the aforementioned split between the aspects which were supposedly related to needs, use values, and the rational condition of economics rationality and the otherwise superfluous elements, ideological values (forged by marketing), and irrational aspects of consumption, fads, and frivolities, and so on.[20] By prescriptively distinguishing a real, rational, and moral dynamic from an artificial, empty, and immoral one, the desire is relegated to an accessory role, to the plane of extra-economic emotions, non-rational subjectivity, and excessive futility, certainly something useless for economic rationality and even less useful to the art of government. However, when doing so, one misses the essence of capitalist mutation and the emergence of sign-money, which worked as the structural condition of capitalism. All in all, the result of this reductionism is just an impotent theoretical approach, lacking enough weapons to confront the real monster.[21] Thus one should have the ears to once again listen to the marginalists, since the innovation of their revolution was operating the *Copernican turn* in relation to the theory of value and, against value objectification, restoring the primacy to subjective desire. According to Goux, "value is no longer value in itself, something inscribed in the heavens of ideas, intrinsic to things or guaranteed by some fixed pattern as universal rule. Above all, value is value-to-me since I am a sensitive and desiring being. Value does not come from the object towards the subject; it comes from the (desiring) subject to then project itself into things, temporarily turning them into valuable objects. It is the object that seems to bend in the face of the demands emanating from the subject—an inversion that may remind us of Cézanne."[22] Desire, that is, sensible passion, comes first: "First and foremost, there is the desire which constitutes the object. This is the basic assumption of such quest. Desire is not desire for a pre-existing object, but the precondition for the existence of the object itself."[23] We cannot find value crystallized in objects.

Value is thus in continuous variation with the subjectivities which, in turn, take part in a desiring process with the objects. Far from establishing an idealist subjectivism, such displacement leads to the subjective production as the mobilizing core of desire and money, which capitalism makes coincide and puts them to work in flows, subsuming the whole process. This occurs both at the macroeconomic level, with the global financial system—which works through trends, gradients, and fluctuations—and the microeconomic level, by means of the continuous overlapping of individual and collective assessments that approach the "value of things" at the exact same time they integrate them into forms of life. Deleuze and Guattari explain this through the concept of "partial object": an "*agencement*" of desire that precedes the very constitution of objects and subjects,[24] which is constantly open for new arrangements, articulations, and entanglements. We have here one more demonstration that "*agencement*" is not an assemblage of fragments, but the articulation of an inseparable artefact of culture and nature.[25]

Desire does not correspond to the mere will of a pure subject, which would then move toward preexisting objects, as a projection. From the very start, desire is the machining of subject and object in the sensible passion that brings them together, which is first in relation to them. Desire exists "in-between" the subject and the object, it precedes the terms of the relation. In other words, desirability is connected to a pre-subjective and intensive human-natural machinery: "The issue is one of *desirability* as an assemblage component: every group desires according to the value of the last receivable object beyond which it would be obliged to change '*agencement*.'" And every "*agencement*" has at least two sides, the machining of bodies or objects, and group enunciation."[26] Transposing it to the analysis of capitalism means that fads, consumer crazes, pop culture explosions, and the like cannot be reduced to the alienation brought by capitalist ideology. There is a *more material and more powerful* desiring process between the sky of values and the earth of labor, which incites a multitude of political, social, and economic effects, which would end up totally dismissed if we were to follow the apothecaries of socialism or the reformers of values with their old litany and moral logic, praising necessity and the concrete aspects of reality. Hence the great merit of the marginalists was identifying how the labor of capitalism rests upon desiring flows that, albeit involving prima facie non-economic aspects (aesthetic, sensual, and ethical elements, for instance), are immediately economic, and they are so *by means of monetary flows*. The mathematization of the capitalist society's logic—which leads Deleuze and Guattari to call it "social axiomatic" in *Anti-Oedipus*—stems from the coincidence between political and libidinal economy, which smoothes the space of relations and transactions that resonates with microeconomic aspects, regimes of the sensible, marketing, cultural production, and other immaterial elements of production.[27]

According to Goux, "the ideological (metaphysical, philosophical) fracture that implies the properly economic notions of *use*, *utility*, *necessity*, and *desire*, is entirely up to the utmost modern and post-modern manifestations of capitalism." The marginalists then introduced innovations to the idea of "useful" that reverberate in the more contemporary organization of capitalism, and this was due to the fact that those theorists were the first to free themselves from the realistic and naturalistic metaphysics on the economic and monetary plane. When replacing the constitutive principle of desire with the principle of absence and all its correspondents—need, scarcity, sufficiency—the 20th-century supporters of the theory of value grossly underestimate the anthropological radicalism of capitalism. For that reason, Goux criticizes Georges Bataille and Jean Baudrillard in particular, but we believe this criticism could apply to almost all critical thinking of the Left, including the workerist (*operaista*) intellectual branch, the *Italian Theory*.[28] Despite the efforts made, during the 1990s, to understand post-Fordism as the new regime of accumulation, those theorists ended up assimilated by the leftist *doxa*, defining the advance of neoliberalism on the welfare state as the central problem.

The critical debate on post-modernism, including its neo and post-Marxist versions, is marked by an interdiction which restricts it to two distinct fields: the neoliberals on one side and the progressives on the other. This normative cleavage operates through a continuous process of silencing every debate that does not cover the consecrated topics of the anti-neoliberalist thinking, which usually represents its nemesis as mere consumer ideology, depoliticizing alienation, and financial hegemony.[29] On his lectures of 1978–1979, published only in 2004, Foucault was the one who broke this barrier. It is no coincidence that he has been considered guilty of the already old accusation of being a "neoliberal philosopher" (or indeed a "new philosopher") since then. But the fact is that the epistemic blackmail of the Left has atrophied their ability to face the challenges presented by the capitalist economy, and they are now incapable of dealing with the transformations that capitalism and currency, both allies, underwent. This obscure counter-hegemony, based on a hygienic conception of the philosophical and political debate, left the whole field open to those who overtly assume neoliberal, orthodox, or monetarist *political* stances—which, as we can see at first inspection, are far from restrained to a single thought. The defeat is twofold: it occurs both in the reality plane, where the tools for dealing with paradoxes and conflicts internal to capitalism are missing, and on theoretical discussion terms, for sheer methodological indigence. The marginalists, however, were far ahead of the theory of value, the use-value moralists, and the back-to-basics naturalistic fairs, even a hundred years before *Anti-Oedipus*, as they identified desire within economy and embraced issues like pleasure, satisfaction, accomplishment, aesthetics, and fruition.

Goux points out that, since 1883, Charles Gide, economist and father of the writer André Gide, "had replaced the term *utility* with *desirability*" to avoid the objective connotation of the former and mark the subjective dimension of desire.[30] Hence, he concludes, the marginalist turn of the 1970s is the point where "we must look for the displacement, the post-modern breach that will give way to the philosophies of desire, which will often and explicitly criticize capitalism (like Deleuze and Guattari or Lyotard), yet without addressing the genealogy which links them to the marginalist rupture."[31] Still, according to Goux, the "emerging political economy did not wait for Thorstein Veblen or the consumer society theorists to free the utility and use value notions from the whole naturalistic ballast and include the gaze of the other in the intensities of desire. This is actually in its birth certificate."[32] Without waiting for Deleuze and Guattari or the culturalist ethnography (Marshal Sahlins), the marginalists have shifted the horizon toward a society of abundance. At which point is the social subject considered satisfied? Is there a limit to consumption? How many units of a given commodity until the desire for such commodity begins to disappear? Most criticisms, even postmodern ones, remain dialectical and reactive in the face of neoliberalism, considering it as if from outside. That way, the critics of the neoclassical economists end up corroborating precisely what they criticize—that is, the illusion of founding a pure economy, built on mathematization and financial econometrics, as if there really were a free market to oppose. On the contrary, Goux says, "there is nothing normative about the economists' notion of 'utility,'"[33] for it is not a substance. What works as a measure of value is satisfaction, which can only be explained within the desiring process, which is immediately social and cultural. Goux highlights a fertile passage from Auguste Walras: "Value is something essentially relative. No doubt there is something absolute behind the relative value, namely, the intensity or emptiness of the last needs met. But that emptiness, which is absolute and not relative, is nothing real or objective, but subjective or personal. They are in us, not on things."[34] This means that the value may be understood as nearly absolute at the precise moment it disappears, the moment it begins to drift, to flow as pure sign. According to the marginalists, this value is subjective, personal, brief, and is determined at that fulfillment point that both suppresses and introduces it as a desert.[35]

The marginalist theory anticipates what, a century later, will be the subject of debates around the postmodern conditions and the capitalist configurations which correspond to it.[36] The poststructuralist effort to mobilize the schematics of structural anthropology, psychoanalysis, and linguistics, leaves nothing but a *déjà vu* feeling in light of marginalism.[37] According to Goux, the topos of unpredictable fluctuations, "liquid" postmodernities, and fragmented subjectivities have already been outlined in the marginalist turn, and picked up the emergence of a "post-bourgeois or postmodern capitalism, which catches the traditional political criticism off-balance and unarmed."[38] The stock exchange model was already a tendency of the flexible regime of accumulation, the cognitivization of labor, and the financialization of life. Within this imaginary, the preference for fluidity marks a displacement of desire, from its connection to properties and possessions toward the abstraction of money. This corresponds to desiring flows escaping from the state and spreading over the surface of the Body without Organs of capitalism. At the same time, this mutation of desire coincides with the mutation of money, when it becomes sign-money, in a self-foundation and self-production cycle, in symbiosis with the "progressive dematerialization of the means of payment and the development of virtual and speculative economy."[39] In other words, the dematerialization of money frustrates, from the start, any possibility for a credible theory of value in modernity. Such theory is only valid in pre-capitalist conditions, as rationalization of the stock-form, but *expires at the exact instant capitalism is born.* Supreme irony: Ricardo develops the theory of value as it lies on its deathbed. He performs a necropsy, not a diagnosis. At that moment, the rationality of measure along with its claim to settle accounts are already over, given the coincidence trend of monetary flows with desiring ones—corollary for an internalized and irredeemable economy of infinite debt (capitalism itself). The neoclassical marginalism anticipates even some aspects of the Nietzschean philosophy in a few years. Nietzsche responds to a nihilist mood, resulting from the death of transcendent values, without resorting to *passive nihilism*, which dismisses the capitalist reality as an illusion of the senses (critique of ideology), or *reactive nihilism*, given to feelings of nostalgia toward the equivalence principle and the "bovine security" of yore (socialism). Instead, he engages it with an *active nihilism*, which can be put into practice within the schizophrenia of desiring/monetary flows: we will call it Schizocene. According to Goux, such dangerous line of thought, deprived of guarantees, presents itself as a "line of flight that disturbs our conception of the relations between reality and virtuality, rationality and irrationality, predictability and randomness." All that converges into a question that seems crucial to us: "How can the aestheticization of political economy (its displacement along with the ambitions and follies of creativity) recover the artistic advances and put them back into the general equivalent logic?"[40] Hence we have two interconnected lines of questioning: on the one hand, the role of the imaginary or, as Goux puts it, the symbolization mode; on the other, the relationship between an economy of desire and the dematerialization of currency. As Nigel Dodd wrote, "money, whenever and wherever used, is not defined by its properties as a material object, but by symbolic qualities generically linked to the ideal of unrestricted empowerment."[41] At the same time, "desire . . . is anticipation, expectation, hope. It is believing in future satisfaction."[42] Hence desire is what creates a load of faith—which is also found in credit, finance, and the stock market: "the symbolic aspects of money as economic instruments of empowerment do not result from the ability of acquiring wealth and property. They are the source of this ability and rely on the abstract power of monetary networks."[43] This rupture is thus up to the challenge of building a theoretical framework of the fiduciary forms of value, since all those forms "attest to the dematerialization of economy."[44] What is at stake is "the emancipation in relation to all moral philosophy,"[45] this poisonous plant whose repeated consumption paralyses the Dionysian creation of other values, of another currency. After all, "desire is the value of value."[46]

Let us trace a parallel lineage: Condillac began by saying that value is not on things, but in the judgment we make of them; Bentham built a moral theory of utilitarianism aimed at maximum collective satisfaction as opposed to the legal model of the French Enlightenment theorists; and Jean Baptiste Say reinforced the idea of a subjective assessment by separating economic value from the more general sphere of religious, civic, and moral values, which would not be economical, thus departing from any normative criterion about what is just or unjust and, therefore, from any theory of justice. Next, with classical political economy, economic science refrains from developing an ethic and asserts the "axiological indifference to political economy" of concepts such as wealth, utility, and value.[47] However, the real turn occurs with both Auguste and Léon Walras, father and son. Auguste Walras overcomes the traditional rhetorical oppositions that define the notion of utility: "Hence there is this difference between moral and political economy, for the former considers useful nothing but the objects that satisfy the needs acknowledged by reason, while the latter grant this same characteristic to all objects that man may desire, whether by motive of self-preservation or due to his passions and whims."[48] Therefore, economic utility overflows the utility framework from a moral point of view and leaves no residue of normativity. Here is another passage from Auguste Walras: "Man does not live on bread alone; he leaves on a bunch of things that, for one reason or another, make his condition more pleasant and agreeable; it is enough that an object somehow contributes to satisfy one of our needs, or provide us with some enjoyment, for it to be useful to us and the economists to declare it so."[49] As Goux points out, younger Walras take up his father's work without any reservation, significantly intensifying what senior Walras did with some scruples. "I claim that things are useful as long as they have some use and allow for one's satisfaction. . . . To us, necessary, useful, enjoyable, and superfluous, all that is just more or less useful."[50] The economist leaves no doubt when he says, "If a substance is procured by a doctor to heal a sick man or by a murderer to poison his family, this is a very important question according to other points of view, but totally indifferent from ours. For us, the substance is useful in both cases, maybe even more so in the latter than the former."[51] As Goux notices, at this point, "there is no ambiguity left in the assertion regarding the axiological indifference of the utility notion." Young Walras thus touches a key aspect of capitalism—that is, its structuring cynicism that stems from the coincidence between sign-money and desirability, an amoral cynicism that cannot be handled by the sheer force of good intensions and social catechisms. His movement is similar to Machiavelli's. Like the theory of power of the latter left no room for considering the negative—as if it were plausible to oppose power and strength with reason and fairness—León Walras is all about taking the bull by the horns, desire against desire, currency against currency, for this is the potentiality grounds mobilized by capitalism. Without resorting to the normativity of moral philosophies, both Machiavelli and L. Walras do not postulate an axiological Outside to counteract the *agencement* of power and desire that are the very material of life. Hence one can say he presents the first "Machiavellian moment" of economic thinking—which Goux prefers to call "cynicism of utility."

In the early 1970s, the shift to post-Fordist regulation, whose milestone was Nixon's statement about the inconvertibility of the dollar, was not a dematerialization process of labor. It was a process of dematerialization of currency. That was the definitive dissolution of value as common and objective measure of wealth. This transition is the same one that leads to the consecration of central banks and the hegemony of global financial institutions. Value dissolves with the emergence of capitalism, since that is when money concludes its futuristic trajectory toward abstraction. What matters for the dynamism of the process is no longer the concrete fortune, the accumulated possessions, but that which is put into circulation, invested, created, and multiplied, that which arises as a pure and fluctuating sign: money, bond, sign-money. With the arrival of capitalism, the material constancy of value, which imprints the sovereign in coins, is replaced by the infinite availability of transaction signs, simple deeds within an axiomatic animated from wall to wall by the warm ocean of desire.[52] We then step into a deregulated regime of signs that refer to other signs, debts on debts, desire on desire, *multiple interpenetrating temporalities*, and so forth. Money, as a Trinitarian formula—unit of calculation, value reserve, and means of exchange—no longer exists. All three functions exploded. The stock form falters and capitalist societies look more and more like primitive societies regarding transversal socialization. But now, unlike societies without state, the whole social tissue is invested with money in terms of an absolute financialization of life. No more rigid codes or ancestral traditions carved in stone, but flows, tides, surges, *Gulf Stream*.

No wonder we witnessed the acceleration of a widespread crisis of political and economic representation in the 1970s—which reached a convulsive level in the 2010s. Only the institutions that were able to conform to the immediacy with which finance governs desire remained: central banks, stock exchanges, rating agencies, hedge funds; investment funds, and joint-stock companies. Capitalism needs no mediation, it simply cuts the middleman. The signifiers have become as empty as the temple: inconvertible currency "that now circulates, runs, whose 'realization' always differs, which provides the monetary function in terms of pure sign."[53] This mutation of the sign leads to the mourning of representation, prelude to a new *nomos* of the earth, marked by the conjunction of powers of global finance, social-digital and algorithmic technologies, and the ultra-connected networks of matter, energy and information (the Great Metabolism). When the whole world is connected within the connective tissue of interaction networks, within the global supply, marketing, and consumption chains, and all operations and payments are carried out without cash, then the monetary deterritorialization saga will have finally experienced its schizophrenic moment. Each and everyone's lives will be fully linked to the electronic signs, the energy beams charged with big data that traverse our bodies, which attest to their activities, relations, projects, and services.[54] The bankarization of life is the new biopolitics frontier, an active nihilism at supersonic speeds, suffering from unbearable pressures.

It is paradoxical that, almost a century and a half after both Walras, Menger, and Jevons, left-wing economists and theorists, on behalf of a prêt-à-porter anti-neoliberalism, keep on postulating the pertinence of a reactive or passive Outside for opposing the monetary and desiring logic. Moreover, in a nostalgic approach, they identify it in the residual Keynesianism of social democracies, in Latin American neo-developmentalism, or in capitalism "with Chinese characteristics." Still more paradoxical is the fact that theorists which happily embraced the pessimism of the negative sometimes seem to be the only ones who escaped this mousetrap, reaffirming life as it is and accepting the desert in which we live. This Sahara of lunar texture, this post-apocalyptic and populated desert, "which is also the founding topos of the marginalist economy,"[55] is the starting point for reinventing desire, desire for desire itself, condition for effective creation. "It is the desert, for the barrenness and deprivation it entails, which enlivens all desires."[56] After all, within scarcity, in times of counterrevolution, the most relevant metamorphosis is not Hegel's owl or Marx's mole, but the Nietzschean camel.[57]

GEORGES BATAILLE: THE CAPITALIST DARK SIDE

Is there a postmodern capitalism? According to classical economics analysis, capitalism is soundly anchored in the theory of value, its rationality principle and intelligibility. Even if those economists may admit that the process of capital develops, over the course of history, in consonance with different modes of social regulation and based on the answers it encounters for the successive crises, for them, renouncing the theory of value would mean placing themselves beyond the framework of analysis and critique of political economy as Smith, Ricardo, and Marx have established. Deepening the crisis internal to capital to its ultimate consequences—as suggested by the eschatological dynamics in the "Fragment on Machines"[58]—would automatically bring us to a post-capitalist stage. In this no-man's land there would be no objective criterion to measure labor and thus no way to endow the capital process with rationality, which, in turn, would annul the possibility of a reformist or socialist political will and rationale. The rule of money would then acquire an entirely political character.

As we have just seen, the logic of money is the main mechanism for bringing off the transition between the Fordist and post-Fordist modes of regulations, when the crisis of economy fully becomes an economy of crisis, and finance gains a unique centrality. The rule of money then acquires an entirely political character. We have also seen how the mutation acts as a hinge within the workings of money, how it works as some sort of corner between the most molar aspects of the capitalist governance—the world financial system, central banks, stock markets—and the molecular ones—the internalizing and individualizing economy of infinite debt. Debt is always solvable when banks are concerned, but it is never the case when it comes to individuals. Speaking on Foucaultian terms, this is the epistemic transition in which currency becomes the main tool of governmentality, the true "police of capital." Within this functioning system, which we call "sign-money," the financial dynamics penetrates deep into the social tissue and reconfigures it, accelerating the metamorphosis of the factory society into the social factory, into the enterprise-society.[59]

To the question regarding the existence of a postmodern capitalism, many left-wing political economists would answer we are experiencing a form of capitalism in which the financial capital assumed an unprecedented centrality, hand in hand with the triumph of the free market and free competition ideology, whose post–Cold War geopolitical expression was the Washington Consensus and Fukuyama's thesis of the end of history in the 1990s. Such an interpretation is still very limited to the analysis of superstructures, as if there were an emanating center of values and ideologies forging an economic materiality. We may find more sharp assessment among those who understand neoliberalism based on an analytics of power, in Foucault's terms. Those analyses tend to value the everyday microtextures, which give substance to neoliberalism as microphysics of power—that is, in the production of the very society as a power plant of values. It does not matter much what is produced in terms of ideology as false consciousness, but how this new productive process, which Foucault describes as the integral method of "governmentality," works. The question concerns the working machine and not some mystical veil covering reality. Thereby, Foucault refers to the set of techniques, devices, and regimes of truth that infiltrate through microeconomics to assert a pervasive principle of efficiency and competitiveness. This results in the creation of the postmodern civil society as the space of the maximization of productive activity, which corresponds to the mutation from the factory regime to the entrepreneurial one, from the mindset of the worker to the entrepreneur rationale, and from the molar logic of full employment to the employability one (precarization, outsourcing of goods and services), when profit is seen from the *rentier* perspective (flexible accumulation). The working relations adopt the service relation model (outsourcing), and the geopolitics of imperialisms give way to the modulated stratification of the *Empire*. In either case, both the ones who address neoliberalism as an ideology and those who associate it to the molecular regime of engenderment, risk converting it into some kind of totality. Whether this whole relates to the geopolitical macrostructure or acts as an omnipresence "from inside," through the social saturation of the molecular fabric, both approaches may end up frustrating any thought regarding struggles and alternatives in the new productive conditions, hindering *ex ante* the deflagration of flight lines.

All that results in a reactive atmosphere of disenchanted analysis, limited to describe the ever more totalizing processes by which we are fabricated, captured, and dispirited. In those cases, there is nothing left but betting on a "backward" way out, a reactive and nostalgic response or an entrenchment bordering on paranoia, on behalf of the last strongholds not yet subsumed by capitalist voracity. Either approach, with more or less sophistication, sticks to saying a rhetorical *no*, armed with the usual anti-neoliberal discourse, as if there was anything left from the old world to defend. Both consist of passive affections which, not by chance, coincide when they decide to choose and support the "lesser evil" of the season. Hence the weak response of a Left which restricts itself to an anti-neoliberal, anti-Right, and anti-capitalist stance, whose militant ethics easily turn into a hysterical hegemonism, always seeking new leaders to postpone the advance of neoliberalism.

Going in the opposite direction, we wish to start from the premise that the same society where even the pores seem under control is also leaking from all sides, that the zenith of capitalism also implies its dawn. It obviously involves more than a decision; we actually must be able to rise to those virtualities. To escape the mousetraps, we approach money as an open field of transformation, with whose dynamics capital formed an alliance, precisely for the fluent potentiality it presents. So, instead of subscribing to the critique of money as an agent of abstraction and alienation—when finance is treated as the liquid vector of speculation in opposition to the solidity of productive labor—we rather criticize the financialization of life in terms of its involvement in an economy driven by *ex nihilo* credit and subjective debt. In order to avoid a similar flattening of the multiple variations of the anti-neoliberal theory, we wish to shed more light on the issues involved in the workings of currency, which touches capitalism in both its molar and molecular regimes of functioning, so we can open the concepts to the strategical thinking *in* postmodernity.

In order to answer the question regarding the existence of postmodern capitalism, J. J. Goux[60] resumed Bataille's anthropologically inflected analyses of political economics in his great book, *La part Maudite*.[61] According to Goux, the French writer made a Copernican turn within political economy in the wake of the marginalist revolution of the 1870s. Instead of defining the dynamo of economic science based on the scarcity problem—that is, on the allocation of production factors aimed at balancing limited goods and unlimited needs—Bataille performs a shift in order to understand the workings of economy from the excess perspective. Besides the closed system of equivalences that the theory of value seeks to correlate, there is an unknown continent that animates the whole economic logic. This dark side, this *part maudite* of the economic phenomenon, often dismissed as a scientific subjectivism, is what truly defines the intensive conditions of the system, according to the incandescent world of desire. The key to a production mode should not be sought in the resolution of the most basic issues of material existence (food, clothing, and dwelling), as Marx and Engels claimed in some of their texts,[62] but "in the blood that spurts from the open chest of victims sacrificed to the sun in an Aztec ritual, in the sumptuous and ruinous feasts offered to the courtiers of Versailles by the monarch of divine right." As Goux concludes, "in all these mad dissipations is found a secret that our restricted economics has covered up and caused to be forgotten."[63]

According to Bataille, the laws of economics based on utilitarian consumption and on expenses guided by a hierarchy of needs, which would condition a science of the virtuous government, become secondary. Within this logic of excess, religion and art come to the fore, for sacred veneration, profane celebrations, and glorious ostentation take precedence in the social organization of desire, the point of maximum intensity of value production. Bataille thus challenges not only rationalist, utilitarian, and productivist formulations of classical political economy but also all economic theories based on the metaphysics of material need, as the one devised in some of Marx and Engels's texts. The driving principles of the economic system are surplus, prodigality (a sort of "anti-production"), the potlach, sacrifice, and waste. This is where the economist should look for the secret of value. Despite Bataille's surplus turn, Goux criticizes the author for not transposing his analysis from the glorious empires and ancient magnificence to capitalism. Bataille would remain too attached to a Puritan conception of capitalism, which goes back to the long tradition that represents the capitalist logic with the image of the greedy, calculating, and ascetic bourgeois. Thus Bataille would end up aligning with the theorizations of capitalism in the line of the sociocultural analysis of Max Weber, for whom the spirit of capitalism found its genealogy in the Puritan Protestant ethics. The world of the Protestant bourgeois would have transposed the religious values of austerity, asceticism, and parsimony to the emergent political economy, favoring the accumulation of capital needed for expanding business. Therefore, the Protestant ethos would be a more adequate spirit to the formation of capitalism, working as a contributing factor to the success of the first national states: Netherlands, France, and England. Goux criticizes what he considers to be Bataille's incomprehension of the real labor of capitalism, when he distinguishes the economies driven by sacred excess (the ancient ones) from desacralized modern economy, based on technique, calculation, and the frugality of savings. According to Goux, this remnant of Weberian flavor can also be found in the criticism of instrumental reason by Frankfurtian Marxists like Horkheimer and Adorno, as well as in most critiques of capitalism as a dehumanizing abstraction and/or a formalist and totalizing rationale.

In order to fill in the gap left by Bataille in his theories of the sacred and the profane transposed to economics, Goux refers to George Gilder, a neoconservative economist linked to the neoliberal American Right. Without aligning to him on the economic level, Goux recognizes his merits for the diagnosis of the properly postmodern dimension of capitalism. In *Wealth and Poverty*,[64] Gilder traces a genealogy of capitalism which connects it to Lévi-Strauss's and Marcel Mauss's ideas on the savage thought (the *pensée sauvage*). Gilder argues that capitalism in the post-modern era would already work within the logic of the potlach, the party, the surplus, thus accomplishing the transition that Bataille left undone. In fact, against the classical purity of Weber's approach (or Benjamin Franklin's), magnificence, subjective desire, irrationality, all that is reworked in the heart of capitalism. Glory and Dionysus are not in the superstructure, but in the infrastructure. The author traces a correspondence between the driving force of desire and the creation of supply, radicalizing Jean-Baptiste Say's thesis that a given supply creates its own demand. The urge for a compensatory return—that is, the need for equivalence and proportionality is unfounded. Indeed, Gilder recognizes that value has no rational or solid basis: "The tale of human life is less the pageant of unfolding rationality and purpose than the saga of desert wandering and brief bounty."[65] Therefore, capitalism is doubly animated by an active nihilism: at one end, the creation of money by banks; at the other, the creation of new products, styles, and markets. We would add yet another aspect to this, namely, the bodies inscribed in the debt economy which asymmetrically fill in the distance between both ends. The contrasts and asymmetries of capitalism are "the realm of dark transcendence where can be found all true light and creativity."[66] Goux sees a successful and sharp diagnosis in Gilder, whose take on the intimate logic of capitalism is far from the theory of value or bourgeois puritanism, and much closer to fashion, art, the art market, the parties' extravagance, the luxury of millionaires, and the apparent chaos of stock exchanges. According to Gilder, as quoted by Goux, "we are witnessing the aesthetization of political economy."[67] No wonder the post-modern entrepreneur has the artist as model: the creative destroyer, the one who take risks and has to surprise and shock society every now and then. Goux does not fail to express his concern when observing that, the moment modern art—anti-bourgeois par excellence—becomes the paradigm for the new version of capitalism, cutting-edge companies start to look more and more like futuristic vanguards. Paradoxically, postmodern capitalism became anti-bourgeois. In *The Coming of Post-Industrial Society*, Daniel Bell has already explained how the transition to a capitalist society of credit and mass consumption would free itself once and for all from the calculating and instrumental rationality of economic puritanism. We thus distance ourselves from the Marxists (both the orthodox and the critics of late capitalism), the classical economists, the Ricardian theory of value, and the *pure and dures* developmentalists. We should thus ask: did the whole plane of immanence then became capitalist? Given the post-Fordist transformation, anticipated in one century by the marginalist economists,[68] would there be no alternative besides populating the desert of creative co-working and post-modern collaborative entrepreneurship? Would submerging in the smooth horizon of the end of history and engaging in a gloomy postmodern irony be the only things left to do?

THE LIVING CURRENCY: BODIES AND MONEY

One of the most significant works for the philosophical theory of currency which, over the 1970s, influenced the productions of J. J. Goux, Foucault, Lyotard, and Deleuze (particularly in *Anti-Oedipus*, with Felix Guattari), was the essay *Living Currency*,[69] by Pierre Klossowski.[70] The ultra-dense text was written as an "extended aphorism," so to speak. According to Klossowski, there is a moralist accusation of capitalism that completely loses sight of the metaphysical subtleties and theological niceties of money. Ascribing the demoralization of life through money to capital means imputing it with a moral force that it does not have, as if capital could devise an ideology so powerful as to manipulate billions of people. Klossowski will short-circuit the split between the infra and superstructure of capital. Against every dialectical scheme, he presents a more complex interpenetration and field of interaction between simulacrum and production, desire and necessity. In capitalism, legislation forbids the direct submission of bodies to an owner. The means of production holder is prohibited from directly requiring sensations from the living body of the workers. Unlike slavery or servitude, in which the *jus primae noctis* prevailed, the capitalist contract is based on wage labor. This is done by interposing money—which appraises consumer goods, wages, and relations by a common measure—within the relationship between employer and employee. People offer their own work to the employer, whereupon they are compensated with payment in cash. According to political liberalism, this mediation provides the workers with a degree of freedom and bodily protection in regard to the owners of capital. In other words, a sphere of human dignity is preserved as it is kept out of commerce, since the worker is forbidden to sell his own body. It would be up to money to stop the perversion of the system, where all cynicisms converge; after all, no matter where the money we carry in our pockets comes from, *pecunia non olet*.

According to Klossowski, the first split between productive goods and superfluous ones revolved around money. As first articulated by political economy scientists, the former support production because, once introduced within the productive process, they are capable of generating more money at the end of said process (the D-D' cycle), whereas the latter—associated with waste, luxury, leisure, and unproductive consumption—would be unable to generate a surplus value and, therefore, multiply money at the end. According to the utilitarian and productivist economic theory, there is a dividing line between useful and efficient goods, on the one hand, and simulacra, on the other. The latter, marked by their merely decorative and amusing functions, would be parasites or accessories of the true cycle of real economy, like lice in the lion's mane. There you have a whole line of critics of the unproductive work, the ones associated with "concrete socialism" and all attempts to reduce the capital process to a rationalization of the production factors toward greater profit, as a productivist machine. As many know, Marx got involved in this original split of economic science when distinguishing, with great emphasis, the productive work/capital from the unproductive one.[71] Klossowski's analysis, however, leads him to conclude that, in fact, useful and superfluous—that is, necessities and fantasies—are just two sides of the same coin, and capitalism can only function within the amalgamation they constitute. Using the orbital bomb and the Callipygian Venus as examples, Klossowski explains how every utensil factory, stemming from efficiency, gives rise to fantasy representation, as well as any factory of simulacra, deriving from fantasy, bring about an affective corporality, a beam of relations and affections. If the bomb instils the terrible fantasy of the nuclear holocaust, the Greco-Roman status is pervaded by the aesthetic enjoyment and erotic fanaticism of its admirers. Both are based on fantasy and, in this sense, according to Klossowski, the Venus is the "smiling face" of the bomb. Money is precisely what establishes a common ground, in which body and representations become so entangled that it is impossible to clearly distinguish them in the face of the voluptuous emotion (Deleuze and Guattari would simply say *desire*) that governs the process. Money is not inert; it is no universal equivalent or mere medium for exchanges. It is living currency indeed, crossed by intensities and passionate polarities, totally implanted with bodies. In one of the book's most unsettling passages, Klossowski says that this is not something that will occur in the future; it is already happening. The shadow economy of desire is a full reality, the animating principle of capitalism. Currency is no medium; this is an attribute of the desiring body, which is a human, all too human, medium. Living currency flows through bodies, objects, and drives according to a libidinal economy. It is alive! Hence both capitalist economic science and its critics happen to be misguided by a fundamental misunderstanding, as they apply a dialectical separation between reality and fantasy, while in the productive process they are all but separated: desire and utility are always comingled. This insufficiently described distinction between desire and utility actually stimulates the apparent and objective movement of capital, which is never just false appearance (a *flatus vocis*, the veil of ideology), but the objective tension between utility and simulacrum, exchange and creation, commensurability and incommensurability, with which capitalism works. Echoing Bataille's formulations concerning pre-capitalism, one must grasp the dark side of capitalist economy: its ghosts, simulacra, desires, and living currency. In short, Klossowski inverts the pair reality-fetish, which is at the heart of metaphysical idealisms and their moral philosophies. We could say this is like glimpsing the first three chapters of *Capital*'s first volume through the kaleidoscopic lens of May 1968.

Still, according to Klossowski, there is no objective value or measure at the heart of money; there is only Eros. Capitalism is actually an erotic-political economy that disperses and condenses Eros in a continuous breathing—sometimes with a slow-paced breath, others almost breathlessly. Eros is the daemon of value. No criticism of political economy will have any effect unless it considers the widespread eroticism which works as its driving shaft. In *Anti-Oedipus*, Deleuze and Guattari took Klossowski's proposition to great lengths in order to match political and libidinal economy, since desire lies within the infrastructure of the economy, not within the superstructure. Political economy = desiring economy. Unlike Marx and Engels in the *German Ideology*, the infrastructure as last instance does not reside in the productivist economy, but actually in the way drives and affections behave. Value and price are inscribed within the depths of emotion, not in a normative rationality of production that could work as a criterion for hierarchizing needs and determining which labor is productive or not. We meet once again marginalist themes like arte art, love, cinema, and fashion. Far from being by-products of real economy, they are all brought back to the engine room of capital. Economic norms cannot be understood as if on a different plane from political, cultural, religious, or legal institutions, all of which are intensively traversed by the desiring economy. If we are to follow the Marxian teaching of entering into the *camera obscura* of production, then we must understand the operating regime of desire in the shadows of inert currency, a "*boudoir* economy" within this field of incommutability, unintelligibility, and the unconscious. We must turn the "there is no free lunch" motto inside out, since our bodies—the productivity of voluptuous emotions—are the ones actually paying for lunch. For the truth is there ain't no such thing as free money. There are no terms of comparison between voluptuous emotion and the apparent movement of exchanges, between wages and equivalents, as there is no common measure between a given amount of work and the sensation logic of the living object. Those distinctions are qualitative and the distance between them is indecomposable. Moreover, there is a fundamental asymmetry in all that, for desire is what prompts value since, without voluptuous emotion, nothing is worth it. This is what constitutes the object from which value can be deduced, in a second moment, according to the subjectivation process that gives rise to the economical agents. In essence, however, nothing has a price until the regime of exchanges, which capitalism put to work in concomitance with the desiring economy, imposes it in order to set value within monetary flows. Capital must stop the process locally in order to establish a price, as if taking a photograph of the moving flow of desire and money intermingled. This does not rebut the existence of a whole monstrous world of perversions and simulacra, the secret life of money before which capitalism must also open the gates, dose the valves, and let the flows gradually go by, under the risk of total devitalization and killing the vampire's victim. The whole capitalist industry was erected around this unholy monster of creative lust, as the little fish that live by the shark's great mouth and, to some extent, *desire* those mortal teeth from where they get their food. Hence this is no voluntary servitude, as if we were witnessing a disease of the will. Instead, this implies a machinic servitude, a given operation of desire when pervaded by capital.

As a consequence, capitalism introduces an absence in an otherwise superabundant reality and organizes the shortage within the social body so that the never-ending multiplication of fantasies of voluptuous emotions cannot directly connect with living bodies. The bodies must be unified as economic subjects full of needs and debts so that money can fill in this gap by indirectly mediating, on the one hand, the power of employers and workers and, on the other, the power of banks and the buying power of wage earners. Regarding the individual, the unit of the self then takes root within a capitalist regime: desire must be created from the organic unit between a *self* that works and a *self* that consumes, so that the value derived from this work will be exchanged for scarce and useful goods. At the same time, capitalism must also fabricate the simulacra so that desire will be invested in the productive organization as a whole, for the entire productive cycle,[72] since the economic infrastructure is nothing more than the libidinal complex of living bodies, upon which everything can have a value. Any other way would mean depleting the process from its potential and interrupting production. Hence there is an internal conflict which is not only social, but psychological. The economic unit of the subject cannot be maintained if, from time to time, he is not led to identify with the deviation and the perversion, inscribing himself within the creative dynamics of desire, otherwise he would devitalize and lower productivity in the overall productive process. Without the superfluous, the excesses, and constant prodigality, the process cannot be concluded, neither from the global nor the individual point of view. This recalcitrant conflict between the self and its desiring dissolution—that is, between self-conservation and death instinct—infiltrates within the organic and psychological unit of the *self*, and collapses it. The social organization will try to accommodate this driving contradiction through dialectical pairs—public and private life, work and affective life, citizenship and consumption, political and economic man, enjoyment at home and sexual perversion, and so forth—upon which capitalism will produce a huge amount of pathologies as well as normalizing cures (and, according to Deleuze and Guattari, psychoanalysis is the most sophisticated one among them). From the drives within the (social or individual) organism arise an internal struggle for or against its unit, between the cage of needs and the birds of fantasy, between passive and active nihilism—two regimens in the desert which capitalism pushes forward. The "drama" of institutional bodies is that they embody the psychological repression of desire at the economy of exchanges level, demanding the fiction of inert currency while having to acknowledge *living currency* and deal with its disparities in order to exploit it. Klossowski illustrates these comings and goings with the image of the vicious circle, a circle of crisis which goes deeper and deeper—to which we could add, in macrostructural terms, the ebb and flow cycles of currency, as in Bernard Schmitt.[73]

In *Living Currency*, Klossowski furthers his analysis by capturing the post-Fordist turn that was beginning to display its signs at the dawn of the 1970s, which shows in the essay as a new industry of suggestion. Referring to advertising, branding, the massification of TV, cinematic production, and fashion, the author explains how the new industry that emerged in the second post-war establishes itself as a monstrous and gigantic simulacrum, benefiting *directly* from the voluptuous emotion for the first time. There is no need for the middleman anymore. The value is now directly captured from celebrity bodies, magazine covers, television advertisement, Hollywood imaginary, Beatlemania, apocalyptic sci-fi, porn industry, pulp comics, and the like, meaning the frenzied production of immediately valuable signs. The living object is not only led to immediately produce *within* capitalism; it is also immediately expropriated and exploited by new modes of biopolitical control (which Foucault extensively discusses under the biopower theme and Deleuze, under the societies of control concept). The constant compromise between fabrication and sublimation of fantasy reaches a point of maximum tension within the post-Fordist institutions—at which moment, according to Deleuze and Guattari, the identity between desire and the socius is taken to the extreme, to the point of virtual rupture. The desiring body is then incorporated into an endless process of iconization, irrevocably relegating the economic distinctions between productive and unproductive work, necessity and desire, objectivity and subjectivity to a faint horizon. It is as if the marginalist revolution left the theoretical field to materialize one hundred years later, as if Léon Walras and Jevons had found their Lenin in the advertisers, designers and pop artists of the great metropolis. How to deal with this incommensurability once it has spread throughout the post-industrial social fabric? How can capitalism reorganize itself when all traditional mediations of political economy have been shattered? The affective turn of the now reformed capitalism destroys once and for all the possibility of mediation through the law of value. Henceforth, capital has to adopt a system of direct command and constant blackmail, a task which will be accomplished by a new regulation based on the ongoing economic insecurity, the widespread subjective debt, and in the modulated governance of the social continuum (workfare). Once the welfare structure of northern countries has been dismantled, the debt economy fully establishes itself between the asymmetric poles of creditor and debtor, in a continuous gradient: on one side, we have the banks, which exist on the condition of being forever savable; on the other, we find an always indebted man, whose obligations usually end up violently, with evictions, exclusions, confiscations, and widespread misery, both on physical and spiritual terms.[74]

Klossowski's *mise en intrigue* on the problem of capitalism allows for the articulation between the economic subject and the affective factory of simulacra: the former, whose apparent rationality is guided by the theory of value and the equivalence principle, is subordinated to the hierarchy of need; the latter, built on the constitutive excess of desire (maximum tension pole), flows through the living bodies. This is the same notion present in *Anti-Oedipus*, in which desiring production amounts to the limit of social production and vice-versa, as two extreme points of distension and contraction, merged within the functioning of capitalism—on the one hand, the inert currency of exchanges, the one mystified by classical political economy and objectivist Marxisms and on the other, the living currency mobilized by voluptuous emotion. It is not that Klossowski dismisses the existence of inert currency and the economy of exchanges, as if they were false appearances. The author just argues that inert currency is only a particular case resulting from the *reduction* of the desire logic. Still, according to him, the existence of an inert pole of money is a structuring aspect for the spread of debt, which will ossify the hierarchy of needs as measurable and comparable variables. It works as the minimum intensity pole within the continuum determined by money from end to end. Living currency, manipulated by the banks and the financial system—the "mutant flow" that appears in Bernard Schmitt's schizophrenic theory—follows the desiring flows. We could say that inert and living currency, which are different in nature, are limited cases of a modulated space of economy, whose gradient between the maximum and minimum intensity of money is given by the credit-debt relation. The lowest potency, or impotence, figures in money as a simple medium for exchange, a reductionist crystallization of value. The highest potency is sign-money—or living currency (Klossowski)—immediately infested with voluptuous emotion and incarnated in living subjects, in desiring bodies. In capitalism, this last pole gains primacy within the set of structural interactions of the system. Klossowski comments on Keynes's insight—who included propensities and emotional variables in his analysis—when the British economist comes to touch this intuition. This manifests in the concept of disutility which, according to Keynes, represents the limit of unintelligibility of economic behavior. Keynes thus saw the phantasmatic constraint of simulacra at the basis of economy that he tried to reparametrize in macroeconomic terms in the face of the transformations brought about by the capitalist crisis during the 1930s, after the New York stock exchange crash and the following economic depression.

Klossowski revisits two authors in order to outline a plan of critical intervention within living currency, thus avoiding to lean on a negative and reactive dialectics, as does "concrete socialism," with its impotent nostalgia. First, the author mentions the utopic formulation of Charles Fourier. Instead of proposing a distributive theory built on the rationalization of value production (as David Ricardo), the French utopist based his revolutionary theory directly on desire. He had an energetic conception of economy, organizing phalansteries, which put common desire in accordance with affinity groups determined by the Law of Attraction. Fourier added plasticity to the production of affection and desires, which would enable a more harmonic re-elaboration of said production, one that revolved around a gift economy striving for a freer game of passions. He is, without a doubt, the first philosopher to underscore that revolution does not arise from reason or necessity, but from desire—never against it. As Klossowski explains, according to Fourier, the weapons of revolution would not be drawn from the unfolding of capital's own dynamic, as it gave birth to its son, the proletariat that would eventually devour it (as Marx argued). Instead, they would emerge from the driving forces that exceed the productive system, as the flows escape social, economic, and political repressions. The problem of such formulation—which gives reason to Marx's criticism of the Utopians—lies in that Fourier does not consider how capitalism itself already works within the irrationalist logic of drives and passions (as theorized by the marginalist economists in the 1870s). There is no other way to bring about a new society than unraveling it from the society which capital itself conditions and explores, unleashing the flames and dangerous tricks that capital, as a magician apprentice, tries to master. Following this line of thinking, *Living Currency* brings Fourier closer to post-modern capitalism, besides circumventing his more idealistic traces.

According to Klossowski, the first to reveal the amalgamation of desire and money within *living currency* was the Marquis de Sade, the last of the Enlightenment men, the one who took the critique of reason to the last consequences, leading it to its drastic culmination. As the *Lumières de la Raison* were going off, Sade found the darkness of desire at the heart of order, the roots of disorder. The revolutionary aristocrat, who died in 1814, after 32 years in prisons and asylums, could see through the new economic institutions of the modern era, which sought to replace the direct exchange of bodies and sensations with the exchange of goods and then proudly declare that the individual subject, now selling "just" his work, was free. This new freedom, announced with so much pomp and circumstance, was nothing less than the newest prostitution. The modern guillotine, which splits the body and head of the worker, holds a dialectical trickery, for it camouflages how money starts to promote the exchange of bodies in the interest of new institutions. The legal defense of human dignity professed by solemn statements implies, in fact, the material and moral prostitution of all within the capitalist regime. According to Sade, capitalism is the realm of generalized prostitution, where each person is summoned to sell oneself or starve to death.[75] With capitalist development, the systematic abuse works through the introduction of a mediation component, namely money. The prostitutes who precede it are not victims of the capitalist system. In reality, the new system generalizes prostitution through a mutation in the logic of money, as it puts all bodies at the service of the production of capital. Faithful to the principles of Enlightenment criticism, Sade enacts the modern logic through the fantasy of its secret societies, gathered from time to time in inaccessible places to perform intricate practices of perversion. Unlike Fourier, Sade proceeds in the anti-utopia style, with no optimism about the possibility of reformulating passions and affections without destroying the bastilles. The societies of libertines enacted fantasies in deserted castles in order to give rise to perversity without any restraint from normative codes except the law of voluptuous emotion. The rituals are meticulously planned and conducted with implacable precision so they can get as close as possible to the subjective essence of desire colonized by capital. The perversion does not introduce the absence, but the excess necessary to the theater of generalized eroticism, where currency settles and then starts catalyzing it.

Following the author's ironic proposal, Juliette—one of the protagonists in Sade's novel by the same name[76]—plunges into prostitution, uncovering the real nature of money. It was not poverty that led her to resort to prostitution; instead, it was wealth that *constrained her* to do so, for now free fun is no fun. After all, what is it worth when money is not involved? Like Juliette's legs, the more the markets open up, and stay wide open, the more corrupted and higher the price. The harmony of the drives within the naïve gratuity logic of Fourier's affective phalansetries is a weak utopianism: with capitalism, nobody wants anything free anymore. Only by selling ourselves can we be worthy! It is not about the mercantilization of eroticism, as the endless moral accusations against prostitution (or capitalism as such) want us to believe. This actually implies the eroticization of the market, which enables an immanent strategy—in Sade's and Klossowski's case, perversion. The monstrosity of desire that Sade reveals in its crudeness is also the potency which begins to unblock and liquefy, within the post–French Revolution at the dawn of capitalist modernity, due to the alliance between capital and money. From his return to Sade, it is worth stressing how Klossowski points to this Sadean perversity as nothing more than the lever for the monstrosity of post-industrial and post-Fordist society, when the very simulacrum becomes the working principle of capitalism. That is to say, this is no longer a matter of closed systems of perversion; we are now dealing with a perversion that overflows from all sides. The hegemonic period of sign-money, its intimate incommensurability, and total financialization are now open to infinite perverted money. We should finally point out that this whole Klossowskian ride through *living currency* unveils a line of flight marked by the "perverted money, which leads to transubstantiation."[77]

NOTES

1. Deleuze and Guattari, "Plateau 13—7000 B.C.: Apparatus of Capture," in *A Thousand Plateaus*, 437–44.

2. See the successful effort at carefully reading this plateau by Guillaume Sibertin-Blanc in the aforementioned *State and Politics*, 19–83.
3. Deleuze and Guattari, *A Thousand Plateaus*, 429.
4. Pierre Clastres, *Society against the State*, trans. Robert Hurley (New York: Uri-zen, 1977).
5. Deleuze and Guattari, *A Thousand Plateaus*, 440.
6. Since *Anti-Oedipus* (chapter 3), the authors call it Urstaat, the Proto-State. For more on that, we once again refer to Sibertin-Blanc's detailed book *State and Politics*.
7. Suzanne de Brunhoff, *L'offre de monnaie: Critique d'un concept* (Paris: Maspero, 1971), 8.
8. Orléan, *The Empire of Value*.
9. Rachline, *De Zéro à Epsilon*, 85.
10. It is not uncommon for anti-Semitic conspiracy theories to closely follow this anti-finance moral condemnation.
11. Negri, *Marx Beyond Marx*.
12. Schmitt, *Monnaie*, 9. Emphasis added.
13. Rachline, *De Zéro à Epsilon*, 86.
14. Orléan, *The Empire of Value*, 106.
15. See, for instance, their take on the collapse of the progressive cycle in Latin America, especially the debacle of the new economic matrix that Brazil tried to implement, as we discussed in chapter 2.
16. Paradoxically, Goux says Foucault was wrong when he said this rupture occurred with Ricardo, since it was Foucault who properly analyzed liberalism in his courses from 1977/78 and 1978/79.
17. Jean-Joseph Goux, *L'art et l'argent: La ruputure moderniste 1860–1920* (Paris: Blusson, 2011).
18. Louis Leroy, "The Exhibition of the Impressionists," Le Charivari, April 25, 1874. The painting at issue was *Impression: Soleil Levant*, 1872, Musée Marmottan Monet, Paris.
19. Michael Hardt and Antonio Negri, *Labor of Dionysus: A Critique of the State-form* (Minneapolis: University of Minnesota Press, 1994).
20. Jean-Joseph Goux, *Frivolité de la valeur: Essa sur l'imaginaire du capitalisme* (Paris: Blusson, 2002), 169.
21. Ibid., 170.
22. Jean-Joseph Goux, *Frivolité de la valeur: Essay sur l'imaginaire du capitalisme* (Paris: Blusson, 2002), 12.
23. Ibid., 54.
24. Deleuze and Guattari, *Anti-Oedipus*, 1–50.
25. See also Giuseppe Cocco, *Mundobraz*, cit., chap. 3.
26. Deleuze and Guattari, *A Thousand Plateaus*, 439. Original emphasis.
27. Ibid., 13. In this sense, the marginalists are somehow capable of integrating criticisms based on a plurality of values. As Elizabeth Anderson put it, "Use, respect, appreciation, consideration, and love are five different ways of valuing things." Cf. *Value in Ethics and Economics* (Cambridge, MA: Harvard University Press, 1995), 10.
28. On the political knot concerning the transition from the mass worker to the social worker, see Gigi Roggero, introduction to *Elogio della militanza: Note su soggettività e composizione di classe* (Derive Approdi, 2016). See also Bruno Cava, "O operaísmo que vem," in *Revista Lugar Comum*, no. 48 (June–August 2016): 5–12.
29. For a detailed discussion on that subject, see chapter 1.
30. Apud Goux, *Frivolité de la valeur*, 158.
31. Ibid., 159.
32. Ibid., 182.
33. Ibid., 14.
34. Léon Walras, *Abrégé des élémentsd'économiepure* (Paris: Pichon, 1938), 153. Apud Goux, *Frivolité de la valeur*, 14–15.
35. Ibid.

36. Marginalists call Deleuze and Guattari's attention for the power of its aberrant logic, not for its economic theory, which they considered "extremely weak," so much so that they see Jevons, for instance, as "a kind of Lewis Carrol of economics." *A Thousand Plateaus*, 437.
37. Indeed, Ferdinand Saussure's linguistic structuralism would already be a silent counter-revolution in the face of marginalism, whose intentions point to an a-significant semiotics, which will later be developed by Louis Hjemslev, Felix Guattari and, more recently, Maurizio Lazzarato. See Maurizio Lazzarato, *Signs and Machines: Capitalism and the Production of Subjectivity*, trans. Joshua David Jordan (Los Angeles, CA: Semiotext[e], 2014).
38. Goux, *Frivolité de la valeur*, 16.
39. Ibid.
40. Ibid.
41. Nigel Dodd, A sociologia do dinheiro, trans. Waldivia Marchiori Portinho (Rio de Janeiro: FGV, 1997), 235. Originally published in English: *The Sociology of Money: Economics, Reason & Contemporary Society* (London: Polity Press, 1994).
42. Goux, *Frivolité de la valeur*, 161.
43. Dodd, *Sociologia do dinheiro*, 245.
44. Ibid., Goux lists the different types of wealth throughout the history of capitalism and political economy: the land-wealth of the Physiocrats, the commodity-wealth of Marx, and the desire-value of Walras.
45. Ibid., 172, 175.
46. Goux, *Frivolité de la valeur*, 64.
47. Ibid., 175.
48. Auguste Walras *apud* ibid., 176.
49. Ibid., 177.
50. Léon Walras, *Éléments d'économie politique pure ou théorie de la richesse sociale* (Paris: Pichon et Durand-Auzias, 1926), *apud* ibid., 177.
51. Ibid., 178.
52. Ibid., 205.
53. Ibid.
54. Rachline, *De Zéro à Epsilon*, 195.
55. Goux, *Frivolité de la valeur*, 59.
56. Ibid., 63.
57. Friedrich Nietzsche, *Thus Spoke Zarathustra*.
58. Karl Marx, "Fragment on Machines," in *Grundrisse*. See also Negri, *Marx Beyond Marx: Lessons on the Grundrisse* (Brooklyn, New York: Autonomedia, 1991). Also from Negri, see *Lessico Marxiano* (Roma: Manifestolibri, 2008), s.v. "Crisi della legge del valore-lavoro."
59. For more on that subject, go to chapter 1.
60. Jean-Joseph Goux, "General Economics and Postmodern Capitalism," *Yale French Studies*, no. 78, On Bataille (1990): 206–24.
61. Georges Bataille, *La part Maudite* (Paris: Les éditions de minuit, 1949).
62. Especially in the first chapter of *The German Ideology*, where they write about the dialectic between productive forces and relations of production.
63. Goux, "General Economics," 206–7.
64. George F. Gilder, *Wealth and Poverty* (New York: Basic Books, 1981).
65. Ibid., 315.
66. Ibid., 309.
67. Goux, "General Economics," 215.
68. See last section of the present chapter.
69. Klossowski, *Living Currency*. It is worth mentioning that the original edition (*La Monnaie Vivante*, 1970) is abundantly filled with erotic-political pictures and drawings.
70. Pierre Klossowki was a French writer who worked with André Gide and was close to surrealists. In 1936 he participated in Geroges Bataille's political project named *Contre-Attaque*. See Michel Surya, "Préface," André Breton, Georges Bataille, *Contre-Attaque*. Uninon de lutte des intellectuels révolutionnaires—1935–1936 (Paris: Ypsilon, 2013).

71. Antonio Negri, *Lessico Marxiano* (Roma: Manifestolibri, 2008), s.v. "Lavoro produttivo/improdutivo."

72. Desire is an inextricable element of the whole productive cycle, from production per se to consumption.

73. See Kerslake, "Marxism and money."

74. Maurizio Lazaratto, *The Making of the Indebted Man: An Essay on the Neoliberal Condition* (New York: Semiotext[e], 2011). From the same author, see also *Il governo dell'uomo indebitato: Saggio sulla condizione neoliberista* (Roma: Derive Approdi, 2012).

75. Decades later, young Marx arrives at the exact same conclusion as Sade: "Prostitution is only a particular expression of the universal prostitution of the worker." Cf. Karl Marx's Economic and Philosophical Manuscripts of 1844.

76. Marquis de Sade, *Juliette, or Vice Amply Rewarded*, 1797.

77. Klossowski, *Living Currency*, 20.

Conclusion

The Body of the Poor

In 1955, after a journey throughout Brazil, Lévi-Strauss published *Tristes Tropiques*, an assortment of philosophical and anthropological reflections written as if a memoir. In his travel log, the author points out the difference between anthropophagy and anthropoemy:

> If we studied societies from the outside, it would be tempting to distinguish two contrasting types: those which practice cannibalism—that is, which regard the absorption of certain individuals possessing dangerous powers as the only means of neutralizing those powers and even of turning them to advantage—and those which, like our own society, adopt what might be called the practice of *anthropoemy* (from the Greek *émein*, to vomit); faced with the same problem, the latter type of society has chosen the opposite solution, which consists in ejecting dangerous individuals from the social body and keeping them temporarily or permanently in isolation, away from all contact with their fellows, in establishments specially intended for this purpose. Most of the societies which we call primitive would regard this custom with profound horror; it would make us, in their eyes, guilty of that same barbarity of which we are inclined to accuse them because of their symmetrically opposite behavior. [1]

The typology of the Americanist anthropologist distinguishes societies depending on whether they operate according to inclusion or exclusion practices. Exclusion marks the civilized societies, which rely on the everlasting reaffirmation of their own interiority—their values, rationalities, and belief systems—as opposed to an outside. Anthropoemy then occurs through the multiplication of dialectical pairs regarding the self and the other, so that the self is always construed as a non-other. According to this logic, primitive is defined as non-civilized, animal or vegetal as non-human, backward or underdeveloped as non-modern, and so on. Encountering the other is then occasion for an inquiry into whether this other is similar to the self and thus endowed with humanity, culture, religion, God. Put simply, if this other has a soul. Inclusion, in turn, relates to the anthropophagic or primitive societies, for whom the vomit logic would seem scandalous, for they do not define themselves by asserting an identity, but through a constant and open relation with all exteriority. Hence otherness is no pretext for probing the nature of the other to determine if it equals the self. It is rather an opportunity to investigate the interference this other may cause in the very self. That is to say, otherness acts as a destabilization and reconfiguration power. Exteriority is thus more important than interiority, in the sense that the society is continuously (re)defined and (un)founded as it devours the relations it establishes with the outside, becoming other in the process. According to Lévi-Strauss, instead of excluding the other from the social body in order to assert their identity and reaffirm a preexistent sameness, anthropophagic societies work by mobilizing the outside in a way that paradoxically constitutes the very social body as a dynamic and driving principle of immanent exteriority.[2]

In August 2012, the graduate program in Arts from the State University of Rio de Janeiro and MAR, the Museum of Art of Rio de Janeiro, organized a seminar called "Vomit and Negation: Anthropoemic Practices in Art and Culture."[3] In retrospect, we could say that the event was embedded in the wave of political and cultural upheaval following the protests against the United Nations Conference on Sustainable Development (Rio+20). The whole turmoil had just taken place in the city, stirring an effervescent flow that would culminate with the June 2013 uprising of the Brazilian multitude, which was particularly intense throughout the metropolitan area of Rio de Janeiro. Embracing the Lévi-Straussian ideas on anthropoemy and anthropophagy as the theme, the organizers' main concern—as they themselves announced—was whether or not postmodern capitalism has become anthropophagic. The point in question was to know if the anthropophagic strategy had simply turned into a mechanism for strengthening the current regime of flexile accumulation. The initial provocation even speaks of an "anthropophagic capitalism" when raising the following question: "In the context of an economic (and social, cultural, political, etc.) system that devours anything and adapts to everything, we ask if this wouldn't be the moment to reconsider anthropoemy as a necessary resistance practice in face of *anthropophagic capitalism*."[4] Such formulation of the problem, asking for a sort of "anthropoemy redux," echoes a particular spirit of the times present in critical theory, shared by a plethora of contemporary thinkers who distrust affirmative theories, filled with flows, becomings, and multiplicities, which would be complicit with the positive and fluid logic of neoliberalism. According to this end-time approach to politics, one would have to recover the power of *no*, negativity, and negative thinking in general. We could mention, for instance,

Slavoj Žižek's criticism of Deleuze and Negri,[5] or Benjamin Noys's accusations against ontological affirmationism.[6] Those formulations are subtly tied to a common theme of the post-1990s Left, which diagnoses a ubiquitous neoliberalism and firmly asserts itself as *anti*-globalization, *anti*-flexibilization of labor, and *anti*-cultural postmodernism. Against the excesses, already assimilated by capitalist accumulation in the post-Fordist era, it would be necessary to interrupt the process and dialectically introduce a sound *no*, even if it means resorting to regressive projects and rebuilding national states, hegemonic parties, and regional geopolitical blocs.

The invitation text of the aforementioned seminar also refers to Oswald de Andrade, whose anthropophagic manifest from 1928,[7] the chef d'ouvre of aberrant Brazilian modernism, called for an ethics of life, meaning an intensive devouring stance in the face of existence. In this seminal text of a movement that crossed generations, the aphorism "I am only interested in what is not mine"[8] (the "law of the anthropophagus") makes room for a set of practices of relational predations, self-variation, and metamorphosis. The Oswaldian manifesto is not limited to prescribing some plain opening to otherness or an ideal of tolerance with difference in a sort of early multiculturalism of *belles âmes*. Such prescriptions would, indeed, be easily accommodated in a post-colonial conception of culture which reduces it to a social commodity and an elitist status marker, an exchange commodity for the political, cultural, and economic upper classes within the universities, art circuits or the creative jet set. As Lévi-Strauss pointed out in *Tristes Tropiques*, Brazilian elites wear culture as if they wore a new dress.[9] Other than that, the Oswaldian program pierces the curtain over the shallow cultural cordiality and touches the primitive unconscious—a "pre-logical mentality," the "palpable existence of life"—which will work as an "anthropophagic vaccine"[10] in the face of the disease of modern Western civilization and the inferiority complexes in the South. The manifesto does not operate by schematics or categories, but by a flood of evocations. The chain of images connects Oswald's poetic intuition to a philosophy of the savage body, in a production—to speak in Deleuzian terms—of a powerful image of thought. The desire that flows through the primitive unconscious is historically repressed by the successive orthopedic operations performed through the colonization of Brazil and the formation of the Brazilian people. From Jesuitical catechesis to Black slavery to the successive waves of industrial proletarianization and urbanization, there were numerous biopolitical technologies to undermine the potencies of the savage body and to mobilize it for work and compliance. Oswald's intuition of the Matriarchy of Pindorama,[11] this mythical image of a naturalist and primitive society without prostitution or penitentiaries,[12] does not correspond to a mere backward-looking approach, a feeling of nostalgia for the indigenous people from the 16th century and the good old days before white men arrived. This was never the case of a romantic ideology trying to revive the image of the Indian as an element of pure and genuine Brazilianness, which will then amalgamate into the Brazilian people concept, minted during the national formation process, built on the idea of three constitutive races.

Oswald is clearly breaking with romantic writers such as José de Alencar as well as with the Indianist nationalists, who, in the 1920s, mystified the image of the Indians for their Wagnerian patriotic cosmogonies, which was as a prelude to the formulations of the 1930s' fascist Right that led straight to the Varguist dictatorship from 1937 to 1945. The Oswaldian Amerindian is neither Rousseau's good savage nor the "Indian dressed up as senator of the Empire . . . featuring in Alencar's operas full of good Portuguese feelings."[13] The nakedness of the naked anthropophagic man is fully clothed in the sense that the body, unlike the Aristotelian tradition in which it is a mere instrument of the soul, is a complex assortment of transitions, a patchwork of affects, relations, and energy. Hence, the myth of the anthropophagic utopia cannot be explained as a revival of a supposed "pre-modern" instinctive reserve, as if the Indians were archetypes of the revolution. Instead, it is a *fully material* actualization of anthropophagic practices, such as horizontal miscegenation and lines of deterritorialization or exodus.[14] The anthropophagic madness has its own right, the cannibal right to oppose the civilizing sickness and rebel against its institutions: the missionary and the church, the master and his land, the merchant of culture and widespread prostitution (capitalism).

We cannot reduce Oswald's experimentation with savage thinking to a mere tradition of cultural critique—which would be aimed against the importers of European canned culture and the reactionary nativists, champions of the Brazilian identity—as if he was just a landmark within the Brazilian cultural history. As Oswald himself had already written in a previous manifesto:[15] "Poetry exists in the facts. The shacks of saffron and ochre in the green of the Favela, under cabralin blue, are aesthetic facts." The primitive unconscious "without complexes"[16] is totally inhabited by rebellious practices and peoples, by the insurgencies from the colonial period, from quilombos and villages. In this sense, it is crossed by thresholds of subversion and powerful psychosocial energies. When Oswald mobilizes the myth against the repression of the unconscious, he does not simply return to the savage, who would lead, in Freudian style, to the de-sublimation of great passions. More than a return, this is about an active oblivion, a pragmatics of de-Oedipalization, a sustained liberation of productive desire. The Dionysian theme present in the anthropophagic manifesto, in its delirious style as well as within the images of idleness, feasting, and free love, does not refer to Freud and the return of the repressed, but to Nietzsche and the transubstantiation of values. We can thus relate the anthropophagic program to the one unfolded by Deleuze and Guattari in Anti-Oedipus. That is to say, it comprises the *production of the corps sauvage* as the limit of neocolonial society, the absolute limit in which becomings and intensities that constitute the Southern unconscious begin to flow freely. It is nothing less than a schizophrenic-anthropophagic revolution which does not go back to the beginning, to a mystified origin, but lies in the future, on the extreme edge of biopower workings. The anthropophagic practices manifest, above all, in neo-primitivism, techno-shamanism, hyper-consumerism, alter-anthropology, as well as in the becoming-Brazil, the Technicized Barbarian,[17] and in all those images pertaining to a theory of subjectivity within underdevelopment, which are pre-logical and post-logical at the same time—that is, if we think like Oswald and go against the eloquent rhetoric of Father Vieira, who believed that "we have never admitted the birth of logic among us."[18] As Oswald himself used to say, instead of an ontology, an odontology. In this sense, his torrential modernist manifesto outlines something like Spinoza's *Ethics* for the tropics. A kind of "Joyeux Tropiques" in which anthropophagy, as Amerindian perspectivism, acts directly on the desiring body and its affections. Transubstantiated into body, the anthropophagic action cannot be framed as a mere culture war, as if it were just a confrontation between belief systems; it can only be approached as a genuine revolutionary force acting through and within bodies. It is not so much a struggle among cultures, but a "culture" of metamorphosis where the body is the protagonist. The "Carahiba revolution" thus implies the mad, nonconformist, *minor* effort against the racist biopower built in the South—and from which Brazil and its national projects arose. We are far from the litanies on acculturation, "ideas out of place," or hegemonic ideology, since anthropophagy implies molecular rebellion, an affective and molecular productive turn, recovering the underdeveloped body in other terms. That is to say, a theory of subjectivity within underdevelopment, an exit from biopower by the means of biopolitics. The anthropophagic delirium, in this sense, is totally historical-political; furthermore, it traces a strategy of institutional creativity beyond any culturalist reduction.

While reporting the difficulties of catechizing the Indians, father Vieira shows he had very well understood how the inconstancy of savage thought implied an active refusal to being dominated for spiritual or physical work. As he manages the marble and the myrtle grove metaphor,[19] the Jesuit realizes that the native soul has its own plasticity. The Brazilian Indians were not like the infidels from the Old World, whose resilience to the catechist chisel was extremely hard to overcome but, once it occurred, their fierce resistance converted to passionate devotion. The Indians simply did not operate under the same logic, meaning adhering or not to a given belief system. Forging stable values which could influence the pragmatics of life seemed an almost impossible task to Vieira. The Indians operated on values as pure positional signs to be devoured within a continuous variation and metamorphosis.[20] Hence, after an initial enthusiasm, the colonial missionaries realized the magnitude of the challenge, since the *yes* of the Indian concealed a second and even bigger *yes*, which proclaimed the living body as the value of values.[21] The indigenous anthropophagy is not the model for an elaborate kind of decisive negation; instead it works as a reference for a double yes, a double affirmation, and a particular type of subjectivity production. Anthropophagists resisted catechesis through a creative reaction by which, when devouring the colonizer's point of view, they changed their own nature, leaving the Jesuits and the *bandeiras* with a truly infernal work of control.

They were like weed, persistent and always renewed. Once trimmed or removed from a place, they would constantly branch or reappear elsewhere. The colonists soon realized that their idea of conversion was chimerical, wherefore the racist biopower mechanisms—which the Lusitanian colonization particularly developed with maximum efficiency and to the most brutal consequences—were introduced even more forcefully. The history of colonization and, later, industrialization—with its lay, positivist, and developmentalist catechesis—bring the anti-memory of biopower resistances along with it, in a continuous reinvention, *from within*, of the racial, territorial, and social modulations. The factory of docile bodies—in which the original project of soul conversion (of the Aristotelian-Thomistic-Jesuitical kind) was forced to become—did not prevent a factory of minor strategies and becomings from surviving in the shadows of the projects for the country, since the capture of the underdeveloped body was never complete. "We were never catechized,"[22] Oswald was saying many decades before Latour outlined the symmetrical anthropology.[23]

But how to deal with postmodern capitalism when its missionaries—such as George Gilder (mentioned by J. J. Goux)—seem to transpose this torrential and baroque plethora into the very language of capital? What about *KorpoBraz* , the savage and desiring body, which is already invested with the process of capital through money, as Klossowski demonstrates in this theory of the *Living Currency* ? Would we be experiencing an anthropophagic capitalism, as the 2012 seminar held in Rio de Janeiro implies?

Suely Rolnik converges with Klossowski's reflections when describing the turn to the post-Fordist flexible regime of accumulation as the era of "globalized pimping."[24] According to the Brazilian schizoanalyst, adherent to Felix Guattari's analysis of integrated world capitalism (IWC),[25] the emergence of neoliberalism corresponds to the moment capitalism ceases to operate conforming to clear limits between form and matter to fully assume moving and blurred contours of subjectivity—in other words, when the production of subjectivity proper to capitalism no longer works by imposing a mold, but through successive modulations, resonances, and energetic variations. Rolnik describes the epochal transition of capitalism as follows: the "rise of the imagination to power is a micropolitical operation that consists in making its potential into the major fuel of an insatiable hypermachine for the production and accumulation of capital—to the point where one can speak of a new working class, which some authors call the 'cognitariat.'" As in Veronica Gago's most recent research,[26] Rolnik analyzes the formation of neoliberalism in terms of a mutation in the regime of sensible textures, which constitutes the very subjects and their affective-relational power—wherein both researches connect to *The Birth of Biopolitics*, Foucault's seminal course from 1978–1979. According to Rolnik,[27] the introduction of the flexible regime of postmodern capitalism was especially fulminating and uncomplicated in Brazil, since the "anthropophagical savoir-faire of the Brazilians gives them a special facility for adapting to the new times."[28]

In fact, Rolnik's reference to the Southern plasticity of subjectivity—or, we could just say, the body—echoes the insights of Jesuit priests, especially Father Vieira, as they notice that the Indians behave like myrtles when facing the colonial invasion. It is as if the whole discussion about the biopolitical, post-Fordist, or postmodern turn was overdue regarding the colonization logic, which had to engender a biopower in order to control the flexible production of subjectivity. What is most important pointing out is that the biopolitical resistances, their deterritorialization lines and subjective exodus, are first in relation to a power that had to infiltrate and parasitize those resistances in order to control from inside. It had to do so because ideological indoctrination, spiritual conversion, or the logic of applying more evolved forms to a supposedly inert and shapeless matter never really worked with the savage body. The fact is that the matter is alive and its lack of shape, monstrously productive, is *living currency*. Rolnik then considers a "Tupinambá capitalism,"[29] in the sense that the "arrival"—the arrival of those who never went—of neoliberalism in the South activated "the worst of this tradition, the basest [low] anthropophagy."[30] The expression low anthropophagy appears in the anthropophagic manifesto of 1928 when Oswald refers to catechism, complexes and the "plague of the so-called cultured and Christianized peoples."[31]

However, as we understand it, the transition from Fordism to post-Fordism in the North, as well as the "arrival" of neoliberalism, cannot be explained by a transition from an anthropoemic and hilemorphic logic (form/content, mold, rigid conception) to an anthropophagic and cannibal one (plasticity, modulation, energetic conception). Anthropophagy cannot be limited to the logic of acculturation, assimilation, class consciousness, belief context, or any culturalist reduction—in short, to a problem of the soul—so one may then assign an anthropophagic quality to the cultural turns and "new spirits of capitalism."[32] The very Jesuit catechists, in a second moment, realized that they could not prioritize the soul, as they did in the Old World. Unlike Rolnik, we understand that anthropophagy is both in the beginning and in the end, as inferior and superior limit of the workings of capital, *since its constitution*. It thus does not imply a "Tupinambá capitalism," but the fact that capitalism has always had to constantly internalize the power of the poor, the *body of the poor*, KorpoBraz. This is also its greatest weakness. In fact, postmodern capitalism is not culturalist, since the exploited bodies are still the body of the poor—and there is nothing more material, productive, and affective than that. *What changed was the body of the poor*. When the becoming-Brazil interferes with the theoretical formulations stemming from Brazil and, more generally, the South, it fosters a displacement from the logic of cultural imperialisms and nationalisms to the logic of biopower and miscegenation. If the pinnacle of anthropophagy construed as a schizoanalysis of Brazil is the savage body—the desert of underdevelopment—then its occupation involves the population by means of miscegenation, flows of migration, and exodus, as well as the refusal of labor, inseparable from biopolitical resistances (the double yes). Hence the relevance, for a biopolitical factory of strategies, of shifting all developmentalist, national formation or proletarianization/awareness topics. This is the only way of rising to the anthropophagic utopia and, at the same time, allowing for a strategical (squizo)analysis of neoliberalism—which is biopower plus sign-money.

Despite the disagreement above, there is another passage regarding Rolnik's convergence between neoliberalism and anthropophagy which we believe is very prolific. When describing the logic of postmodern capitalism, she outlines the simultaneous functioning of two regimes: "This kind of pimping of the creative force is what has been transforming the planet into a gigantic marketplace, expanding at an exponential rate, either by including its inhabitants as hyperactive zombies or by excluding them as human trash."[33] This does not entail a typology, but a topology—that is, two poles of different natures that are constantly interacting within the overall workings of the system. That is to say, inclusion and exclusion are two operations that, combined, stratify the capitalism of post-Fordist regulation all the way. There is nothing more distant from a *flat horizon* than the multiple modulations and gradients of force that neoliberalism governs in the continuum—which, as we have seen, Foucault called governmentality. As the author herself explains, "[i]n fact, those two opposing poles are interdependent fruits of the same logic; all our destinies unfold between them" through "the opposite and complementary figures of luxury and trash subjectivity."

In order to reclaim the anthropoemy-anthropophagy pair, we could reframe the issue in terms of low and high anthropoemy as well as low and high anthropophagy, considering that capitalism must remain working within the range between the upper and lower limit in each case. When analyzing the production of subjectivity or bodies in neoliberalism, we thus have two ever-present machinic processes (exclusion and inclusion) in a field of constant interaction. The original typology of Lévi-Strauss then evolves into a topology. This is no longer a matter of anthropoemy *versus* anthropophagy, but each one as two sides of the same coin,[34] the double regime whereby biopower plus money, hence postmodern capitalism, works. As the anthropophagic thought, postmodern capitalism is both at the beginning and at the end. In this sense, the capitalist crisis is also a crisis capitalism—that is, the ability to maintain the interval, to extend the limits of accumulation and, at the same time, prevent anthropophagy from completely imploding the building of exploitation and domination. Thinking in terms of Klossowski's concepts, resistance arises from two perversions, one for each regime: anthropoemy as creative refusal—an escape that looks for a weapon as well as an exodus which also populates the desert where the people are absent—and anthropophagy as it crosses the biopower modulations—intensification beyond the limit of crisis, its qualitative variation. We could still add that, under those terms, anthropoemy and anthropophagy do not represent a dialectics but from the point of view of capital, for the double perversion/affirmation has the ability to put an end to the dialectics by eliminating the relationship. Immanent exteriority—the savage body of the South, the Body of the Poor—is the absolute limit of such topology, the living source which can neither be suppressed nor fully controlled at the risk of devitalizing the parasitic process. The Left is constantly defeated by neoliberalist forces because it still has not abandoned the aspirations of the first generation of Jesuits—that is, imposing a form (state-form, development, redeeming industry) on the living matter of underdevelopment. Those forces, in turn, caught up with the second wave of less optimistic priests like Vieira and recognized its inner power. So instead of trying to mold the living matter, they let the flows run and then govern within the very flexibility. This is what neoliberalism does better; it is its expertise and raison d'être. Defusing the collective consciousness from the oedipal-narcissistic fixations of a form-lover Left is undoubtedly part of the schizo-Amerindian effort of the Oswaldian program. However, overcoming neoliberalism implies a second movement, a second affirmation, namely accelerating the process, devising a theory of subjectivity within underdevelopment, connecting to the biopolitical factory of strategies, following the lines of what we call the becoming-Brazil.

After we have traced a genealogy of the biopower mechanisms in Latin America and outlined a positive program of minor politics, we engaged then in the task of writing an Ethics of the anthropophagic unconscious amidst the swirling flow of the 2013 uprising throughout Brazilian metropolises. Deleuze and Guattari claimed that the whole theoretical endeavor of *Anti-Oedipus* was an attempt to adapt the Ethics of Immanence to the new flexible coordinates of capitalism, to the point of saying that the Spinozian substance reappeared in the fluid and mad concept of Body without Organs (BwO). And each of us have our own BwO, we all make our own BwO. In this sense, we could say that the present book operates as an interaction between the displacement toward biopower with the political and ethical formulations of the figure of the body of the poor. Always with the struggles and for the struggles, our whole work of dismantling the great theoretical-political buildings of developmentalism, of positivist rationalism, of progressivism, of the promises of national formation, of the redemptive proletarianization, and of the popular culture project of a statist and statolatrous Left suffering from a guilty conscience thus culminates in the (de)definition and (un)foundation of Brazil. Brazilianization is useless from the point of view of transformation, since we, like Oswald, are not interested in the Indian dressed up as senator of the Empire, the "torch-bearing Indian," meaning the Brazilianization forged by nationalisms, including the people of national-popular consciousness. We then arrive at *our* BwO, at the desert, but an intensely populated desert, crossed by miscegenation gradients, the mobile thresholds of migrations, by biopolitics, "against all catechesis," the assertion of the *Body of the Poor*.

NOTES

1. Claude Lévi-Strauss, *Tristes Tropiques* (Penguin UK, 2011). Original emphasis.
2. Commenting on this passage, the Brazilian anthropologist Eduardo Viveiros de Castro explains the tremendous asymmetry between the behavior of the Portuguese discoverers who arrived in the New World and the indigenous people who came across them for the first time. While the Europeans immediately threw themselves into a theological-political discussion as to whether the Indians had souls and, therefore, were human like them, the Indians, in turn, seriously wondered if the Europeans were gods or mythical beings, i.e., if they were endowed with a different body from theirs. While the European anthropology stood by principles of identity and similitude, putting the soul at the heart of the matter, the Amerindian alter-anthropology was guided by difference, having the body as the main issue—that is, the set of abilities and powers capable of affecting and being affected. See Eduardo Viveiros de Castro, "O mármore e a murta: sobre a inconstância da alma selvagem," in *A inconstância da alma selvagem* (São Paulo: Cosac Naify, 2014).
3. See "Em parceria com Museu de Arte do Rio (MAR), UERJ realiza seminário sobre antropoemia na arte," Portal Fator Brasil, August 21, 2012, accessed January 31, 2017, http://www.revistafatorbrasil.com.br/ver_noticia.php?not=214068.
4. Ibid.
5. Slavoj Žižek, *Organs without Bodies* (New York: Routledge, 2016). From the same author, see also *Living in the End Times* (London: Verso, 2011).
6. Benjamin Noys, *The Persistence of the Negative: A Critique of Contemporary Continental Theory* (Edinburgh: Edinburgh University Press, 2010).
7. Oswald de Andrade, "Manifesto Antropófago," *Revista de Antropofagia* 1, no. 1 (May 1928). For an English version of the text, see "Anthropophagic Manifesto," trans. Maria do Carmo Zanini, Sibila, March 24, 2009, accessed January 31, 2017, http://sibila.com.br/english/anthropophagic-manifesto/2686.
8. Ibid.
9. Lévi-Strauss, *Tristes Tropiques*.
10. Andrade, "Manifesto Antropofágico."
11. The term means "land of palm trees" in Tupi and, by a stretch of language, stands for Brazil, whose coastal area was covered with such vegetation.
12. Andrade, "Manifesto Antropofágico."
13. Ibid.
14. As we have described in chapter 2.
15. Oswald de Andrade, "Manifesto Pau-Brasil," *Correio da Manhã*, March 1924. For an English version of the text, see "Manifesto of Pau-Brasil Poetry," trans. Stella M. de Sá Rego,

Latin American Literary Review 14, no. 27 (Jan.–Jun. 1986): 184–87, accessed January 31, 2017, http://www.jstor.org/stable/20119419.

16. Andrade, "Manifesto Antropofágico."
17. Ibid.
18. Ibid.
19. Padre Antônio Vieira, "Sermão vigésimo sétimo do Rosário," in *Padre Antônio Vieira: Essencial*, ed. Alfredo Bosi (São Paulo: Penguin, 2011), 532–33.
20. Castro, "O mármore e a murta."
21. On the double affirmation, see the Christ of Nietzsche's *Antichrist*, trans. H. L. Mencken (New York: Cosimo, 2005).
22. Andrade, "Manifesto Antropofágico."
23. Bruno Latour, *Jamais fomos modernos: Ensaio de Antropologia Simétrica* (Rio de Janeiro: Editora 34, 2005).
24. Suely Rolnik, "The Geopolitics of Pimping," trans. Brian Holmes, European Institute for Progressive Cultural Policies, October 2006, accessed January 15, 2017, http://eipcp.net/transversal/1106/rolnik/en.
25. Feliz Guattari and Suely Rolnik, *Cartografias do desejo* (Petrópolis: Vozes, 1986).
26. Veronica Gago, *La rázon neoliberal: Economias barrocas y pragmatica popular* (Madrid: Traficantes de Sueños, 2015). For a more thorough discussion on neoliberalism, see chapters 1 and 2.
27. Suely Rolnik, "Anthropophagic Subjectivity," in 24a Bienal de São Paulo: Arte Contemporânea Brasileira: Um e/entre outro/s, ed. Paulo Herkenhoff and Adriano Pedrosa (São Paulo: Fundação Bienal, 1998), 128–47. Bilingual Portuguese/English edition. Republished in Portuguese as "Subjetividade Antropofágica," in Daniel Lins, comp., *Razão Nômade* (Rio de Janeiro, 2005: Forense Universitária).
28. Rolnik, "The Geopolitics of Pimping."
29. Speaking at a lecture. See also "Zombie Anthropophagy," in Ivet Curlin and Natasa Ilic eds., *Collective Creativity* (Kassel: Kunsthalle Fridericianum, 2005). Bilingual German/English edition. Published in a shorter version in French as "Anthropophagie Zombie," in *Mouvement: L'indiscipline des Arts Visuels*, no. 36–37 (Paris, Sept.–Dec. 2005): 56–58.
30. Rolnik, "The Geopolitics of Pimping."
31. Andrade, "Manifesto Antropofágico."
32. See Jean-Luc Boltansky and Éve Chiappello, *Le nouvel esprit du capitalism* (Paris: Gallimard 1997). By the way, it would be more appropriate to speak of "animal spirits" of capitalism. See Matteo Pasquinelli, *Animal Spirits: The Bestiary of the Commons* (Rotterdam: NAi Publishers/Institute of Network Cultures, 2008).
33. Rolnik, "The Geopolitics of Pimping."
34. The same goes for the problem of currency, which is both sign-money—the monstrous *living currency* that is also credit-money, governed by central banks—and exchange-currency that, as in Viveiro de Castro's critique of *Anti-Oedipus* regarding Marcel Mauss, is also double-sided—the measure-exchange (low anthropoemy) and metamorphosis-exchange (high anthropoemy). We could outline a similar redistribution of terms regarding horizontal and vertical miscegenation, as well as for the lines of territorialization and deterritorialization.

Bibliography

Acosta, Alberto. *O bem viver: Uma oportunidade para imaginar outros mundos*. Translated by Tadeu Breda. São Paulo: Autonomia Literária/Elefante, 2016.

Agamben, Giorgio. *Homo Sacer: il potere sovrano e la nuda vita*. Torino: Boringhieri, 1995.

Aglietta, Michel and Bai, Guo. *La voie chinoise: capitalisme et empire*. Paris: Odile Jacob, 2012.

Alencar, José de. *Cartas a favor da escravidão*, [1867]. comp. Tâmis Parron. São Paulo: Hedra, 2008.

Alencastro, Luiz Felipe de. *O trato dos viventes*. São Paulo: Companhia das Letras, 2000.

Altamira, César. *Os marxismos do novo século*, [2006] translated by Leonora Corsini Rio de Janeiro: Civilização Brasileira, 2008.

Amar, André. "Essai psychanlitique sur l'argent," *Revue française de psychanalyse* 20, no. 3: 332–344, 1956. http://bsf.spp.asso.fr/index.php?lvl=notice_display&id=13586.

Amato, Massimo and Fantacci, Luca. *Fine della finanza: da dove viene la crisi e come si può pensare di uscirne*, 2nd ed. Roma: Donzelli, 2012.

Anderson, Elizabeth. *Value in Ethics and Economics*. Cambridge, MA: Harvard University Press, 1995.

Andrade, Oswald de. "Manifesto Antropófago," *Revista de Antropofagia* 1, no. 1[1928]. English version "Anthropophagic Manifesto," translated by Maria do Carmo Zanini, Sibila, March 24, 2009, accessed January 31, 2017, http://sibila.com.br/english/anthropophagic-manifesto/2686.

———. "Manifesto Pau-Brasil," *Correio da Manhã* [1938]. For an English version of the text, see "Manifesto of Pau-Brasil Poetry," translated by Stella M. de Sá Rego, *Latin American Literary Review* 14, no. 27. Jan.-Jun. 1986: 184–87, accessed January 31, 2017, http://www.jstor.org/stable/20119419.

———. "Aqui foi o sul que venceu," in *Ponta de Lança*,[1949]. 5th ed., Obras Completas São Paulo: Editora Globo, 2004. 104–12. Originally published in *O Estado de São Paulo*, September 2.

Arantes, Paulo. *O novo tempo do mundo – e outros estudos sobre a era da emergência* São Paulo: Boitempo, 2014..

Arrighi, Giovanni. *Il lungo XX secolo: Denaro, potere e le origini del nostro tempo*. Translated by Mauro Di Meglio. Milano: Il Saggiatore, 2014. English edition *The Long Twentieth Century: Money, Power and the Origins of our Time*. New York: Verso, 1994.

———. *Adam Smith in Beijing: Lineages of the Twenty-first Century*. London-New York: Verso, 2007. Spanish edition: *Adam Smith en Pekín: Orígenes y fundamentos del siglo XXI*. Translated by Juanmari Madariaga. Madrid: Akal, 2007.

Arrighi, Giovanni and Zhang, Lu. "Dopo il neoliberismo: Il nuovo ruolo del Sud nel mondo," in Giovanni Arrighi. *Capitalismo e dis)ordine mondiale*. Roma: Manifestolibri, 2010.
Augé, Marc. *Non-lieux: Introduction à une anthropologie de la surmodernité*. Paris: Seuil, 1992.
Balakrishnan, Gopal ed. *Debating Empire*. London: Verso, 2003.
Balazs, Étienne. *Chinese Civilization and Bureaucracy: Variations on a Theme*. New Haven: Yale University Press, 1962.
———. *La bureaucratie celeste*. Paris, Gallimard, 1968.
Bataille, Georges. *La part maudite*. Paris: Minuit, 1949.
Beck, Ulrich. *What is globalization.* Cambridge: Polity Press, 1999.
Becker, Jasper. *Hungry Ghost: China's Secret Famine*. London: John Murray Publishers, 1996.
Belluzo, Luiz Gonzaga. "Foucault e o neoliberalismo," *Carta Capital*, May 24, 2013. accessed February 2016, http://www.cartacapital.com.br/revista/750/foucault-e-o-neoliberalismo-1420.html.
Bianco, Lucien. *La récidive: Révolution russe, révolution chinoise*. Paris: Gallimard, 2014.
Bielschowsky, Ricardo. *Pensamento econômico brasileiro: O ciclo ideológico do desenvolvimentismo*. Rio de Janeiro: Contraponto, 1995.
Billeter, Jean-François. *Chine Trois Fois Muette*. Paris: Allia, 2000.
———. *Contre François Jullien*. Paris: Allia, 2006.
Blackstone, Brian and Walker, Marcus. "Draghi driblou alemães na defesa do Euro," *Valor Econômico*, October 9, 2012.
Bloch, Marc. *Esquisse d'une histoire monétaire de l'Europe*. Paris: Armand Colin, 1954.
Boltanski, Luc and Chiapello, Ève. *Le nouvel esprit du capitalisme*, Paris: Gallimard, 1999. English edition: *The new spirit of capitalism*. London: Verso, 2007
Boutang, Yann Moulier. *De l'esclavage au salariat: Économie historique du salariat bridé* Paris: PUF, 1998.
———. *Le Capitalisme Cognitif: La Nouvelle Grande Transformation*. Paris: Amsterdam, 2007. English edition: *Cognitive Capitalism*. Translated by Ed Emery, London: Polity Press, 2012.
Braudel, Fernand. *La dynamique du capitalisme historique*. Paris: Arthaud, 1985.
Brunhoff, Suzanne de. *L'offre de monnaie: Critique d'un concept.* Paris: Maspero, 1971.
Cardoso, Fernando H. "Introduction," *O Mito da Marginalidade*, by Janice Perlman, Rio de Janeiro: Paz e Terra, 1977.
———. *Capitalismo e Escravidão no Brasil meridional: O negro na sociedade escravocrata do Rio Grande Do Sul*. Rio de Janeiro: Civilização Brasileira, 2003 [1962].
Cardoso, Fernando H. and Faletto, Enzo. *Dependency and Development in Latin America*. Berkeley: University of California Press, 1979.
Carneiro, Lucianne. "Os europeus ainda não olharam para exemplos como o do Brasil," *O Globo*, October 25, 2012.
Castro, Eduardo Viveiros de. *A inconstância da alma selvagem* São Paulo: Cosac Naify, 2002.
———. "The relative native" translated by Julia Sauma and Martin Holbraad, *Journal of Ethnographic Theory HAU,* 8, no. 8 2013.
———. *Cannibal Metaphysics*. Minneapolis: Univocal Publishing, 2014 [2009].
———. "Immanence and Fear: Stranger-Events and Subjects in Amazonia," *The relative native* London: Hau Books, 2015.
Castro-Gómez, Santiago. "Ciências sociais, violência espistêmica e o problema da 'invenção do outro,'" in Edgardo Lander ed. *A colonialidade do saber*. Eurocentrismo e ciências sociais. Perspectivas latino-americanas, São Paulo: Clacso, 2005.
Cava, Bruno. "La misteriosa curva della retta lulista," Commonware website. December 3, 2014, http://commonware.org/index.php/cloe/521-curva-retta-lulista.
———. "Pueden los gobiernos progresistas sobrevivir al proprio éxito?," *Lobo Suelto!* blog. January 6, 2016, accessed August 2016. http://anarquiacoronada.blogspot.com.br/2016/01/pueden-los-gobiernos-progresistas.html.
———. "O operaísmo que vem," in *Revista Lugar Comum*, no. 48 June-August 2016.

Cava, Bruno and Cocco, Giuseppe. "Vogliamo tutto! Le giornate di giugno in Brasile: la costituzione selvaggia della moltitudine del lavoro metropolitano," *EuroNomade*, October 21, 2013, http://www.euronomade.info/?p=173.

Cava, Bruno and Cocco, Giuseppe. Eds. *Amanhã vai ser maior: O levante da multidão no ano que não terminou*. São Paulo: AnnaBlume, 2014.

Chakrabarty, Dipesh. *Provincializing Europe: Postcolonial Thought and Historical Difference*. Princeton: Princeton University Press, 2000.

Chamayou, Grégoire. *A Theory of the Drone*. New York: New Press, 2015.

Chang, Jung and Halliday. Jon. *Mao: The Unknown Story*. New York, Knopf, 2005.

Chavez, Gonzalez. "El Dilema Neoliberal De Los Subsidios," *El Día*, January 29, 2012, accessed March 2016, http://www.eldia.com.bo/index.php?cat=162.

China Labour Bulletin, January 7, 2016, http://www.clb.org.hk/en/content/strikes-and-protests-china%E2%80%99s-workers-soar-record-heights-2015.

Clastres, Pierre. *Society against the State*. Translated by Robert Hurley, New York: Uri-zen, 1977.

Coppola, Frances. "China's Black Monday Signals the End of Its Growth Cycle," *Forbes*, August 24, 2015, accessed July 2016, http://www.forbes.com/sites/francescoppola/2015/08/24/chinas-black-monday-signals-the-end-of-its-growth-cycle/#69f895f219b2.

Cocco, Giuseppe. "A Cidade policêntrica e o trabalho da multidão," *Lugar Comum* 10, April 2000, 61–89.

———. "Democracia e Socialismo na era da subsunção real: A construção do comum," in *O mundo real: Socialismo na era pós-neoliberal*, by Tarso Genro Porto Alegre: L&PM, 2008. 55–91.

———. *Mundobraz*: *o devir-Brasil do mundo e o devir-mundo do Brasil*. Rio de Janeiro: Record, 2009.

———. "Não Existe Amor No Brasil Maior," *Le Monde Diplomatique Brasil*, May 2013.

———. *KorpoBraz: por uma Política dos Corpos*, Rio de Janeiro: Mauad X, 2014.

———. Ed.: "*The Insurgent Multitude in Brazil*," *The South Atlantic Quaterly* 113, no. 4 Fall 2014.

———. "Brésil, 13 ans après. Entretien avec Giuseppe Cocco," Vacarme, June 5, 2016, http://www.vacarme.org/mot992.html.

———. "La encrucijada de la izquierda brasileña," Nueva Sociedad—Democracia y politica en America Latina, August 2016, accessed August 2016, http://nuso.org/articulo/la-encrucijada-de-la-izquierda-brasilena/.

Cocco, Giuseppe and Negri. Antonio, *Glob(AL): Biopoder e luta em uma América Latina globalizada*. Translated by Eliana Aguiar Rio De Janeiro: Record, 2005.

Comaroff, Jean and John L. *Theory from the South. Or, how Euro-America Is Evolving toward Africa*, London: Paradigm, 2012.

Cooper, Melinda. "The Law of the Household: Foucault, Neoliberalism, and the Iranian Revolution," in *The Government of Life: Foucault, Biopolitics, and Neoliberalism*, ed. Vanessa Lemm and Miguel Vatter. New York: Fordham University Press, 2014.

Corsini, Leonora. "As migrações e o trabalho da resistência," *Revista Lugar Comum*, no. 19–20, July 2009: 211–21, accessed March 2016, http://uninomade.net/wp-content/files_mf/113003120838As migrações e o trabalho da resistência - Leonora Corsini .pdf.

Correa, Rafael. "A Europa comete os mesmos erros que nós," *Le Monde Diplomatique Brasil*, December 2013.

Culp, Andrew. *Dark Deleuze*. Minneapolis: University of Minnesota Press, 2016.

Cunha, Euclides da. *Backlands: The Canudos Campaign*. Translated by Elizabeth Lowe, New York: Penguin Classics, 2010 [1902].

Dantas, Monica. *Revoltas. motins, revoluções: Homens livres pobres e libertos no Brasil do século XIX*. São Paulo: Alameda, 2011.

Dardot, Pierre and Laval, Christian. *The New Way of the World: On Neoliberal Society*. London: Verso, 2013.

Deleuze, Gilles and Guattari, Félix. *Anti-Oedipus: Capitalism and Schizophrenia*. Minneapolis: University of Minnesota Press, 1983 [1972].

———. *A Thousand Plateaus: Capitalism and Schizophrenia*. Translated by Brian Massumi, London/Minneapolis: Minnesota Press, 1987 [1980].

———. *What is philosophy?* New York: Columbia University Press: 1994 [1991].

Deleuze, Gilles. *Foucault*. London: A&C Black, 2006 [1986].

———. "Postscript on the Societies of Control," *October* 59 (Winter 1992 [1990]): 3–7, http://www.jstor.org/stable/778828.

Deutsch, Emeric. *L'argent: Données et débats. Actes du XXVIIIe colloque des intellectuels juifs de langue française*. Paris, Denoël: 1989.

Dikötter, Frank. *Mao's Great Famine: The History of China's Most Devasting Catastrophe, 1958–1962*, London: Bloombury, 2010.

Dodd, Nigel. *A sociologia do dinheiro*. Translated by Waldivia Marchiori Portinho, Rio de Janeiro: FGV, 1997. 235. English Edition: *The Sociology of Money: Economics, Reason & Contemporary Society*. London: Polity Press, 1994.

Dostoyevsky, Fyodor. *The house of the dead*. Mineola, New York: Courier Corporation, 2014.

Douzinas, Costas and Žižek, Slavoj Ed.. *The idea of communism*. London: Verso, 2010.

Droulers, Martine. *Brésil, une géohistoire* Paris: Presses Universitaires de France, 2001.

Errejón, Galván Iñigo. "La lucha por la hegemonía durante el primer gobierno del MAS en Bolivia 2006–2009): un análisis discursivo" PhD diss., Universidad Complutense De Madrid, 2012. accessed August 2015. Available at http://eprints.ucm.es/14574/.

Fiori, José Luis. "O Sistema interestatal capitalista no início do século XXI," in *O mito do colapso do poder americano*, ed. José Luis Fiori, Carlos Medeiros, and Franklin Serrano Rio De Janeiro: Record, 2009.

———. "Sobre o desenvolvimento chinês – II," *Valor Econômico*, March 27, 2013.

———. "Sobre o desenvolvimento chinês I." In *História, estratégia e desenvolvimento: para uma geopolítica do capitalismo*. São Paulo: Boitempo Editorial, 2015.

———. "Dúvidas e certezas americanas," *Valor Econômico*, May 5, 2015.

Foucault, Michel. *Discipline and Punish: The Birth of the Prison*. New York: Vintage Books, 1995 [1975].

———. "Nietzsche, Genealogy, History," in *Language, Counter-memory, Practice: Selected Essays and Interviews*, ed. Donald F. Bouchard Ithaca, NY: Cornell University Press, 1977.

———. *Security, Territory, Population: Lectures at the Collège de France, 1978–79* New York: Palgrave Macmillan, 2007 [2004].

———. *The Birth of Biopolitics: Lectures at the Collège de France, 1978–79*. Translated by Burshell, Graham (New York: Palgrave Macmillan, 2008 [2004]).

———. "Le Philosophe masqué," in *Dits et écrits* II, 1976–1988. Bedroom, Paris: Gallimard, 2001 [1994]. 923–929.

———. "A governamentalidade," in *Estratégia, poder-saber*, by Michel Foucault, comp. Manoel Barros da Motta, 3rd ed., vol. IV, Ditos e Escritos Rio de Janeiro: Forense Universitária, 2012. 281–305.

———. *Le gouvernement de soi et des autres* (1982–1983). Paris: Gallimard, 2008.

Fukuyama, Francis. *The End of History and the Last Man*. New York: Free Press, 1992.

Gonzalez, Pedro Blas. *Ortega's the "Revolt of the Masses" and the Triumph of the New Man*, New York: Algora Pub, 2007.

Fumagalli, Andrea and Mezzadra, Sandro eds.. *Crisis in the Global Economy: Financial Markets, Social Struggles, and New Political Scenarios*. London: Semiotexte. 2010.

Fumagalli, Andrea and Mezzadra, Sandro eds. *La Moneta Nell'impero*, ed. Andrea Fumagalli *et al.* Roma: Ombre Corte, 2002.

Furtado, Celso. *A economia Brasileira: Contribuição à análise do seu desenvolvimento*. Rio de Janeiro: A Noite, 1954.

———. *Uma economia dependente*. Rio de Janeiro: A Noite, 1956.

———. *Essencial Celso Furtado*. São Paulo: Companhia das Letras, 2013.

Gago,Veronica. *La rázon neoliberal: Economias barrocas y pragmatica popular*. Madrid: Traficantes de Sueños, 2015.

Galbraith, John Kenneth. *La moneta: da dove viene e dove va*. Translated by Capriolo Milan: Mondadori, 1976. Originally published in English: *Money: Whence It Came, Where It Went* 1975.

Galvão, Alexander *et al.*, ed. *Capitalismo cognitivo: Trabalho, redes e inovação*. Translated by Eliana Aguiar, Rio de Janeiro: DP&A, 2003.
Genovese, Eugene. *From Rebellion to Revolution: Afro-American Slave Revolts in the Making of the Modern World*. Baton Rouge: Louisiana State University Press, 1980.
Gilder, George F. *Wealth and Poverty*. New York: Basic Books, 1981.
Gilroy, Paul. *The Black Atlantic*. Modernity and Double Consciousness. Cambridge: MIT Press, 1993. Brazilian edition: *O Atlântico negro: Modernidade e dupla consciência*. Translated by Cid Knipel, Moreira São Paulo: Ed. 34, 2001.
Gluckman, Max. *Custom and Conflict in Africa*. Glencoe, 111: Free Press, 1959.
Goody, Jack. *La raison graphique: la domestication de la pensée sauvage*. Translated by Jean Bazin and Alban Bensa. Paris: Minuit, 1979. Originally published in English: *The Domestication of the Savage Mind*, Cambridge: Cambridge University Press, 1977.
———. *L'Orient en Occident*. Translated by Pierre-Anotine Fabre Cambridge: Seuil, 1999. in English: *The East in the West*. Cambridge: Cambridge University Press, 1996.
———. *The Theft of History*. Cambridge: Cambridge University Press, 2006. French translated by Fabienne Durand Bogaert, *Le vol de l'histoire* (Paris: Gallimard/Folio, 2010).
Goux, Jean-Joseph. "General Economics and Postmodern Capitalism," *Yale French Studies*, no. 78. On Bataille 1990.
———. *Frivolité de la valeur: Essa sur l'imaginaire du capitalisme*. Paris: Blusson, 2002.
———. *L'art et l'argent: La ruputure moderniste 1860–1920*. Paris: Blusson, 2011.
Graeber, David. *Debt, the First 5,000 Years*, 2011. Melville House, Brooklyn & London, 2014.
Guattari, Félix and Rolnik, Suely. *Cartografias do desejo*. Petrópolis: Vozes, 1986.
Gutiérrez, Bernardo. "Ecuador pone rumbo a la economía del bien común," *#CÓDIGOABIERTO_CC* author's blog. Accessed March 2016, http://codigo-abierto.cc/ecuador-pone-rumbo-a-la-economia-del-bien-comun/.
Hardt, Michael, and Negri, Antonio. *Labor of Dionysus: A Critique of the State-form*. Minneapolis: University of Minnesota Press, 1994.
———. Antonio. *Empire*. Cambridge: Harvard University Press, 2000.
———. *Multitude: War and Democracy in the Age of Empire*. New York: Penguin, 2004.
———. *Commonwealth*. Cambridge, MA: The Belknap Press of Harvard University Press, 2009.
Hardt, Michael. "Porto Alegre: Today's Bandung?," *New Left Review* 14, March/April 2002. accessed June 2016, https://newleftreview.org/II/14/michael-hardt-porto-alegre-today-s-bandung.
Harvey, David. *The New Imperialism*. Oxford: Oxford University Press: 2003.
———. *A Brief History of Neoliberalism*. New York: Oxford Press University Press, 2007.
Hegel, Friedrich G. *Phenomenology of Spirit*. Translated by A. V. Miller, Oxford: Clarendon Press, 1977 [1807].
Hirschman, Albert. "The Political Economy of Import-Substituting Industrialization in Latin America," *The Quarterly Journal of Economics* 82, no. 1, 1968.
Hirschman, Albert. *Essays in Trespassing: Economics to Politics and Beyond*. Cambridge: Cambridge University Press, 1991.
Holston, James. *Insurgent Citizenship: Disjunctions of Democracy and Modernity in Brazil*. Princeton: Princeton University Press, 2008.
Huntington, Samuel. *The Clash of Civilizations and the Remaking of World Order*. New York: Simon & Chuster, 1996.
Jevons, Stanley. *Theory of Political Economy* (Lausanne: Rouge, 1896).
Joxe, Alain. *Les guerres de l'empire global*. Paris: La Découverte. 2012.
Jullien, François. *Procès ou création: Une introduction à la pensée des lettrés chinois*. Paris: Éditions Du Seuil, 1989.
Jullien, François. *Chemin Faisant: connaître la Chine, relancer la philosophie. Réplique à ****. Paris: Seuil, 2007.
Jullien, François. *The Book of Beginnings*. Translated by Jody Glasdding, New Haven, CT: Yale University Press, 2015.
Jullien, François. *Les Transformations Silencieuses Chantiers, I*. Paris: Grasset-Poche, 2009. 32–33.

Kantorowicz, Ernest. *The King's Two Bodies: Studies in Mediaevel Political Theology*. New Jersey: Princeton University Press, 1997 [1957].

Karmy Bolton, Rodrigo. *Ensayos sobre razón imperial y mundo árabe contemporáneo*. Santiago de Chile: Lom Ediciones, 2016.

Kerslake, Christian. "Marxism and money in Deleuze and Guattari's *Capitalism and Schizophrenia*: On the Conflict between the Theories of Suzanne de Brunhoff and Bernard Schmitt," *Pahrresia*, no. 22. http://www.parrhesiajournal.org/parrhesia22/parrhesia22_kerslake.pdf.

Keynes, John Maynard. *The General Theory of Employment, Interest, and Money*. Macmillan and Co., 1964 [1936].

———. *The Collected Writings of John Maynard Keynes. Volume 1: The Economic Consequences of the Peace*, London: Macmillan, 1971 [1919].

———. "Economic possibilities for our grandchildren 1930," in *The Collected Writings of John Maynard Keynes. Volume IX: Essays in Persuasion* Basingstoke: Palgrave Macmillan, 2010.

———. "Auri Sacra Fames 1930," in *The Collected Writings of John Maynard Keynes*. Cit.

Kissinger, Henry. *On China*. New York: Penguin Press, 2011.

Klein, Naomi, *No Logo, Talking Aim at the Brand Bullies*. New York: Picador, 1999.

———. *No is Not Enough: Defeating the New Shock Politics*, London: Allen Lane, 2017.

Klossowski, Pierre. *Living Currency*. London: Bloomsbury Academic, 2017 [1970].

Krugman, Paul. "Lecciones argentinas para Europa," *El País*, May 16, 2012.

Lagasnerie, Geoffroy de. *La dernière leçon de Michel Foucault: Sur le néolibéralisme, la théorie et la politique*. Paris: Fayard, 2012.

Landes, David. *The Rise of Capitalism*. New York: Macmillan, 1966.

Lane, Frederic C. "Venetian Bankers, 1496–1533: A Study of the Early Stages of Deposit Banking," *The Journal of Political Economy* 45, no. 2. April 1937: 187–206.

Lander, Edgardo ed. *A colonialidade do saber*. Eurocentrismo e ciências sociais. Perspectivas latino-americanas, São Paulo: Clacso, 2005.

Latour, Bruno. *We Have Never Been Modern*. Cambridge, MA: Harvard University Press, 1993.

Lazzarato, Maurizio and Negri, Antonio. *Trabalho imaterial: Formas de vida e produção de subjetividade*. Translated by Mônica Jesus Rio de Janeiro: DP&A, 2001.

Lazzarato, Maurizio. *The Making of the Indebted Man: An Essay on the Neoliberal Condition*, Los Angeles: Semiotext(e)/Smart Art, 2012.

———. *Signs and Machines: Capitalism and the Production of Subjectivity*, London: Semiotext(e). 2014.

Levinas, Emmanuel. "Le débat russo-chinois et la dialectique," *Esprit*, no. 10 1960: 1622–24.

Levinas, Emannuel. "Sur l'esprit de Génève," *Esprit*, no. 1, 1956.

Levinas, Emmanuel. *Les imprévus de l'histoire*. Paris: Fata Morgana, 1994.

Lévi-Strauss, Claude. *Tristes Tropiques*. Penguin UK, 2011 [1955].

Leroy, Louis. "The Exhibition of the Impressionists," Le Charivari, April 25, 1874. The painting at issue was *Impression: Soleil Levant*, 1872, Musée Marmottan Monet, Paris.

Lima, Flavia and Watanabe, Marta. "'Preço alto das commodities é como cocaina,' diz Pettis," *Valor Econômico*, September 25, 2015. Interview with Michael Pettis.

Lin-Chan, Houa Hua Linchan. *Les années rouges*. Paris: Seuil, 1987.

Linebaugh, Peter and Rediker, Marcus. *The Many-headed Hydra: Sailors, Slaves, Commoners, and the Hidden History of the Revolutionary Atlantic*. Boston: Beacon Press, 2000.

Linera, Alvaro. "Indianismo y marxismo: El desencuentro de dos razones revolucionarias," *Revista Donataria* 2, March/April 2005.

Llosa, Mario Vargas. "El regreso del idiota," *Foreign Policy* 160, May/June 2013.

Lo, Sonny. *Hong Kong's Indigenous Democracy: Origins, Evolution and Contentions*, Basingstoke: Palgrave Macmillan, 2015.

Lula da Silva, Luíz Inácio *Carta ao povo brasileiro*, May 22, 2012. *Apud* "Leia íntegra da carta de Lula para acalmar o mercado financeiro," *Folha Online*, June 24, 2002, accessed February 2016, http://www1.folha.uol.com.br/folha/brasil/ult96u33908.shtml.

Lyotard, Jean-François. *La condition postmoderne*, Paris: Minuit, 1979.

Machado, Decio. "Ecuador: la Nueva Disputa por el Poder," *La brecha,* 18.08.2017, http://uninomade.net/tenda/3443/.

Magnier, Mark. "The Wall Street Journal, Shenzhen," *Valor Econômico*, December 22, 2015.

Mao Tse-Tung, "Talks on questions of philosophy 1964," in *Mao: On Practice and Contradiction*. London-New York: Verso, 2007.

Marazzi, Christian. "Money in the World of Crisis: The New Basis of Capitalist Power," in *Zerowork: Political Material* 2, October 1976.

Marshall, Tim. *Prisoners of Geography*, London:CPI Group. 2015.

Marx, Karl. "Reply to Milkhailovski" November 1877, in Karl Marx and Friedrich Engels, *Basic Writings on Politics and Philosophy*. Garden City, N.Y.: Doubleday, 1959.

———. "Preface to the First Edition," in *Capital: A Critique of Political Economy*, by Karl Marx, vol. 1 Hardmondworth, 1990.

———. "Chapter XXXIII—The Modern Theory of Colonisation," in *Capital: A Critique of Political Economy*, vol. 1. North Chelmsford, MA: Courier Corporation, 2011.

Marx, Karl and Engels, Friedrich "Manifesto of the Communist Party," *Marxist Internet Arquive,* accessed July 7, 2016. https://www.marxists.org/archive/marx/works/1848/communist-manifesto/ch01.htm#007.

Massoumi, Brian. *Organs without Bodies*. New York: Routledge, 2016.

McLeod, Alasdair. "China's 1929 Moment," Goldmoney/Wealth, November 30, 2001, accessed June 2016. https://wealth.goldmoney.com/research/goldmoney-insights/chinas-1929-moment.

Means, Gardiner and Berle, Adolf Augustus. *The Modern Corporation & Private Property*. New Brunswick: Transactions Publishers, 1968.

Mèlés, Baptiste. "Experience and Subjectivity: François Jullien and Jean François Billeter," *Personality and Subjectivity, East and West*, Dec 2009, Clermont-Ferrand, France.

Mendes, Alexandre Fabiano. *Conflito e produção de subjetividade no capitalismo contemporâneo,* PhD diss., Universidade do Estado do Rio de Janeiro, 2012.

Mendes, Alexandre, Cava, Bruno and Cocco, Giuseppe. "Foucault e os desafios para a América Latina," *Le Monde Diplomatique Brasil*, June 8, 2015. http://www.diplomatique.org.br/artigo.php?id=1897.

Merleau-Ponty, Maurice. *Les aventures de la dialectique*, Paris: Gallimard-Folio, 1955.

Mezzadra, Sandro. *Diritto di fuga*. Migrazioni, cittadinanzam globalizzazione, Verona: Ombre corte, 2006.

———. *Nei cantieri marxiani: Il soggetto e la sua produzione*. La Talpa: Manifestolibri, 2014.

Mezzadra, Sandro, Reid, Julian and Samaddar, Raṇabir. *The Biopolitics of Development: Reading Michel Foucault in the Postcolonial Present*. New Delhi: Springer, 2013.

Mezzadra, Sandro and Neilson, Brett. *Border as Method, or the Multiplication of Labor*, Durham: Duke University Press, 2013.

Mignolo, Walter D. *Local histories/Global designs: Coloniality, Subaltern Knowledges and Border Thinking*, Princeton: Princeton University Press, 2000.

Milios, John, Dimoulis, Dimitri and George Economakis, *Karl Marx and the Classics: An Essay on Value, Crises and the Capitalist Mode of Production*. Burlington: Ashgate Publishing, 2002.

Mintz, Sidney W. *Sweetness and Power: The Place of Sugar in Modern History*. New York: Penguin, 1985.

Moreno, Camila. *O Brasil made in China: Para pensar as reconfigurações do capitalismo contemporâneo*. São Paulo: Fundação Rosa Luxemburgo, 2015.

Moura e Souza, Marcos de. "China não vai se contrapor aos EUA na AL, diz analista," *Valor Econômico*, June 6, 2015.

Muñoz, Gerardo. "Notas sobre el agotamiento del ciclo progresista latinoamericano," *Infrapolitical Deconstruction Collective*, October 29, 2015, accessed August 2016, https://infrapolitica.wordpress.com/2015/10/29/notas-sobre-el-agotamiento-del-ciclo-progresista-latinoamericano-gerardo-munoz.

Murphy, Timothy S. *Antonio Negri: Modernity and the Multitude*, Cambridge: Polity, 2012.

Mustapha, Abdul-Karim. "O Paradigma das duas fronteiras do Brasil," translated by Patrícia Farias, *Revista Lugar Comum*, no. 11 August 2000): 15–22, accessed March 2016, http://

uninomade.net/wp-content/files_mf/111712120505O paradigma das duas fronteiras no Brasil - Abdul-Karim Mustapha.pdf.
Nathan, Andrew. "Jade and Plastic," *London Review of Books* 27, no. 22 (November 17, 2005): 10–13.
Needham, Joseph. *La Science chinoise et l'Occident Le grand titrage*. Translated by Eugène Jacob, Paris: Seuil, 1973. Originally published in 1969.
Negri, Antonio. *Potere Costituente. Le alternative della modernità*, Milan: Sugar&co: 1992.
———. *Revolution Retrieved: Writings on Marx, Keynes, Capitalist Crisis, and New Social Subjects 1967–83*. London: Red Notes, 1988.
———. *Marx Beyond Marx: Lessons on the Grundrisse* Brooklyn, New York: Autonomedia, 1991 [1979].
———. *Spinoza et nous*. Paris: Galilée, 2010.
———. *Dalla fabbrica alla metropoli: Saggi politici* Roma: Datanews, 2008.
———. *Lessico Marxiano*. Roma: Manifestolibri, 2008. s.v. "Lavoro produttivo/improdutivo."
Negri, Antonio and Cocco, Giuseppe. *Glob(AL): Biopoder e luta em uma América Latina globalizada*. Translated byEliana Aguiar. Rio De Janeiro: Record, 2005.
Neri, Marcelo. A *Nova Classe Média*: o lado brilhante dos pobres, Rio de Janeiro: Saraiva-FGV, 2012).
Neto, Olímpio Cruz. "FHC e o fim da Era Vargas," *OLÍMPIO CRUZ NETO* website. September 1, 2009. https://olicruz.me/2009/09/01/fhc-e-o-fim-da-era-vargas/.
Neuman, William. "Turnabout in Bolivia as Economy Rises From Instability," *The New York Times*, February 16, 2014, accessed March 2016, http://www.nytimes.com/2014/02/17/world/americas/turnabout-in-bolivia-as-economy-rises-from-instability.html.
Noys, Benjamin. *The Persistence of the Negative: A Critique of Contemporary Continental Theory*. Edinburgh: Edinburgh University Press, 2010.
Nietzsche, Friedrich. *Antichrist*. Translated by H. L. Mencken. New York: Cosimo, 2005.
Nobre, Marcos. *Imobilismo em movimento*. São Paulo: Companhia das Letras, 2013.
OECD/CAF/ECLAC, *Latin American Economic Outlook 2016: Towards a New Partnership with China*. Paris: OECD Publishing, December 2015.
Oliveira, Francisco de. "Crítica da razão dualista" [1972], in *Crítica da razão dualista/O ornitorrinco*. São Paulo: Boitempo, 2013. 25–119.
Ong, Aihwa. "Self-fashioning Shanghainese: dancing across Spheres of Value," in *Privatizing China: Socialism from Afar*, ed. Li Zhang and Aihwa Ong. Cornell University Press: Ithaca-London, 2008.
Orléan, André. *The Empire of Value: A New Foundation for Economics*. Translated by M. B. DeBevoise. Cambridge: The MIT Press, 2014.
Orrell, David and Chlupaty, Roman. *The Evolution of Money*. New York: Columbia University Press, New York: 2016.
Ortíz, Alejandra Santillana. "Cracks in Correísmo?" *Jacobin*, 08.14.2015, https://www.jacobinmag.com/2015/08/correa-ecuador-pink-tide-protests-general-strike/.
Pasquinelli, Matteo. *Animal Spirits: The Bestiary of the Commons*. Rotterdam: NAi Publishers/Institute of Network Cultures, 2008.
Pasquinelli, Matteo ed., *Alleys of Your Mind: Augmented Intelligence and Its Traumas*. Lüneburg: Meson Press, 2015. Available at http://matteopasquinelli.com/?p=2872.
Parsons, Talcot. *Societies: Evolutionnary and Comparative Perspective*. New Jersey: Prentice-Hall Incorporated, 1966.
Pessoa, Fernando. "Mensagem" accessed April 2016. Translated by by João Manuel Mimoso, http://www.inverso.pt/Mensagem/MarPortugues/Padrao.htm.
Phillips, Tom. "Families of Tiananmen Square Victims Accuse Beijing of Three Decades of 'White Terror,'" *The Guardian*, June 1, 2016.
Piore, Michael J. and Sabel, Charles F. *The Second Industrial Divide*. New York: Basic Books, 1984.
Polanyi, Karl. *The Great Transformation: The Political and Economic Origins of Our Time*, 2nd ed. Boston, MA: Beacon Press, 2001 [1944].
Pomar, Wladimir. *O enigma chinês*, 2nd ed. São Paulo: Fundação Perseu Abramo, 2015.

Prebisch, Raúl. *El desarrollo económico de la América Latina y algunos de sus principales problemas*. Santiago: CEPAL, 1949.

———. *Problemas teóricos y prácticos del crecimiento económico*, Santiago: CEPAL, 1973 [1951].

Preciado, Beatriz. *Texto Junkie*. Translated by Bruce Benderson from French. New York, Feminist Press, 2013 [2008].

Quijano, Anibal. "Colonialidade do Poder, eurocentrismo e América Latina," in Edgardo Lander ed. *A colonialidade do saber*. Eurocentrismo e ciências sociais. Perspectivas latino-americanas, São Paulo: Clacso, 2005.

Rachline, François. *De Zéro à Epsilon*: vers une nouvelle théorie de l'economie. Paris: First, 1991.

———. *Que l'argent soit: Capitalisme et alchimie de l'avenir* Paris: Calmann- Lévy, 1993.

Ramonet, Ignacio. "La pensée unique," *Le Monde Diplomatique*, January 1, 1995, see https://www.monde-diplomatique.fr/1995/01/RAMONET/6069.

Read, Jason. *The Micro-politics of Capital: Marx and the Prehistory of the Present* Albany: State University of New York Press, 2003.

Rey, Jean-Michel. *Le temps du crédit.* Paris: Desclée de Brouwer, 2002.

Ribeiro, Darcy. *O Povo Brasileiro: A formação e o sentido do Brasil*. São Paulo: Companhia Das Letras, 1995. 95–98.

Rocha, Glauber. "Prefácio 80 1980," in *Revolução do Cinema Novo*. São Paulo: Cosac Naify, 2004.

Roggero, Gigi. "Introduction" to *Elogio della militanza: Note su soggettività e composizione di classe*. Derive Approdi, 2016.

Rolnik, Suely. "Anthropophagic Subjectivity," in 24 Bienal de São Paulo. Paulo Herkennhoff and Adriano Pedrosa. *Arte Contemporânea Brasileira. Um e.entre outro/s*. São Paulo: Fundação Bienal, 1998. 128–47. Portuguese/English Edition.

———. "Zombie Anthropophagy," in Ivet Curlin and Natasa Ilic eds., *Collective Creativity*, Kassel: Kunsthalle Fridericianum, 2005. Bilingual German/English edition. Published in a shorter version in French as "Anthropophagie Zombie," in *Mouvement: L'indiscipline des Arts Visuels*, no. 36–37 (Paris, Sept.-Dec. 2005): 56–58.

———. "The Geopolitics of Pimping," translated by Brian Holmes, European Institute for Progressive Cultural Policies, October 2006, accessed January 15, 2017, http://eipcp.net/transversal/1106/rolnik/en.

Roover, "Aux origines d'une technique intellectuelle. La formation et l'expansion de la comptabilité à partie double," *Annales d'histoire économique et sociale* 9, no. 45 (May 31, 1931): 270–298, http://www.jstor.org/stable/27574484.

Rosenberg, Nathan, and Birdzell, L. E., *How the West Grew Rich*. New York: Basic Books, 1986. Translated by André Charpentier, *Comment l'occident s'est enrichi* Paris: Fayard, 1989.

Ross, Andrew. "China's next cultural revolution?" in *Nice Work if You Can Get It: Life and Labor in Precarious Times*. New York Univ. Press: New York-London, 2009.

Ross, Andrew. *Nice Work if You Can Get It: Life and Labor in Precarious Times*. New York, New York University Press, 2009. 53–76.

Rosset, Clément. *L'anti-nature*, 5th ed. Paris: PUF, 1973. 18.

Sade, Marquis de. *Juliette, or Vice Amply Rewarded*, 1797.

Said, Edward W. *Orientalism*. New York: Pantheon, 1978. Rancière, Jacques. "Foreword to the english edition," in *Althusser's Lesson*. Translated by Emiliano Battista. New York: Continuum, 2011.

Santos, Boaventura de Souza. "Brasil: A Grande Divisão," *Carta Maior*, November, 5, 2014, accessed June 2016, http://www.cartamaior.com.br/?/Coluna/Brasil-A-Grande-Divisao/32167.

Sanyal, Kalyan. *Rethinking Capitalist Development: Primitive Accumulation, Governmentality and Post-Colonial Capitalism*. New York: Routledge, 2014 [2007].

Schavelzon, Salvador. *Plurinacionalidad y vivir bien/buen vivir: Dos conceptos leídos desde Bolivia y Ecuador post-constituyentes*. Quito: Abya Yala/CLACSO, 2015.

Schavelzon, Salvador. "El fin del relato progresista en América Latina," *La Razón*, June 21, 2015, accessed August 2016, http://www.la-razon.com/index.php?_url=/suplementos/animal_politico/fin-relato-progresista-America-Latina_0_2292970735.html.

Schmitt, Carl. *Der Nomos der Erde im Völkerrecht des Jus Publicum Europaeum*. Berlin: Dunker & Humblot, 1974.

Schmitt, Carl. *Land and Sea*. Translated by Simona Draghici, Washington, D.C.: Plutarch Press, 1997.

Schmitt, Bernard. *Monnaie, Salaires et profits*. Albeuve: Castela, 1975 [1966].

Sibertin-Blanc, Guillaume. *State and Politics: Deleuze and Guattari on Marx*. Translated by Ames Hodges, Los Angeles: Semiotext(e)/Smart Art, 20126. 112.

Simmel, Georg. *Philosophie des Geldes*. Translated by Sabine Cornille and Philippe Ivernel, Paris: PUF, 1987 [1900].

Simons, Henry Calvert. *A Positive Program for Laissez Faire: Some Proposals for a Liberal Economic Policy*. Chicago: University of Chicago Press, 1949.

Singer, André. Os *Sentidos do Lulismo: Reforma Gradual e Pacto Conservador*. São Paulo: Companhia das Letras, 2012.

Sloterdijk, Peter. *In the World Interior of Capital*. Cambridge: Polity Press, 2013.

Soares, Luiz Eduardo. *Rio de Janeiro: Extreme City*. Translated by Anthony Doyle, Penguin Books, 2016.

Soja, Edward. *Postmodern Geographies*, London-New York: Verso, 1989.

Sombart, Werner. *Economic Life in the Modern Age*, ed. Nico Stehr and Reiner Grundmann, New Brunswick: Transaction Publishers, 2001.

Soto, Hernando de. *El otro sendero—la revolución informal*, Lima: Instituto Libertad y Democracia, 1986. *The Other Path: The Economic Answer to Terrorism*, New York: Harper & Row Publisher, 1989.

Soto, Hernando de. *The Mystery of Capital*. New York: Basic Books, 2000.

Souza, Jessé. *Os batalhadores brasileiros: Nova classe média ou nova classe trabalhadora?*, 2nd ed. Belo Horizonte: UFMG, 2012 [2010].

Starhawk, *Making It Real: Initiation Instructions: Seattle '99*, PDF, December 1, 1999. Available at http://starhawk.org/Activism/activism%20writings/1999-WTO%20Articles/Making%20It%20Real-Day%202.pdf.

Stefanoni, Pablo. "La utopia coreana en los andes," *Rebelión*, author's blog. July 18, 2013, accessed March 2016, http://www.rebelion.org/noticia.php?id=171279.

———. "La lulización de la izquierda latinoamericana," *Le Monde Diplomatique*, September 2014, accessed March 2016, http://www.eldiplo.org/notas-web/la-lulizacion-de-la-izquierda-latinoamericana.

———. "La lulización de la izquierda latinoamericana," *Le Monde Diplomatique*, September 2014, http://www.eldiplo.org/notas-web/la-lulizacion-de-la-izquierda-latinoamericana.

Sugihara, Kaoru. "The East Asian Path of Development: A Long-Term Perspective," in *The resurgence of East Asia*, Arrighi, Hamashita, and Selden. New York: Routledge, 2003.

Szaniecki, Barbara. "Marée Amarildo: amour et art de la multitude," *Multitudes*. Paris: 56. 2014.

———. *Outros Monstros Possíveis, Disforme Contemporâneo e Design Encarnado*. São Paulo: AnnaBlume, 2014.

Tible, Jean et al., "Dossiê: A potência dos pobres," *Lugar Comum* 40, December 2013.

Toledo, Fabricio. "Fuga como resistência: A pobreza criando excedentes," *Revista Lugar Comum*, no. 31–32 (February 2012): 103–13, accessed March 2016, http://uninomade.net/wp-content/files_mf/110410120852Fuga como resistência - a pobreza criando excedentes - Fabrício Toledo de Souza.pdf.

Tomich, Dale. *Through the Prism of Slavery: Labor, Capital, and World Economy*. Lanham: Rowman & Littlefield, 2004.

Torre, Carlos de la. "El tecnopopulismo de Rafael Correa: ¿Es compatible el carisma con la tecnocracia?" *Latin American Research Review* 48, no. 1 (Spring 2013): 24–43, accessed March 2016, https://muse.jhu.edu/article/504727.

Tronti, Tronti. *Operai e capitale*, 2nd ed. Torino: Einaudi, 1970.

Uehara, Alexandre Ratsuo. "China's Foreign Policy and Bilateral Relations with Brazil," in *China and Brazil: Challenges and Opportunities*, ed. Leia da Costa Ferreira and José Augusto Albuquerque, São Paulo: Anna Blume, 2013.

Valdés, Juan Gabriel. *Pinhochet's Economists*, The Chicago School in Chile. Cambridge: Cambridge University Press, 1995.

Velho, Otávio. "A ideologia da miscigenação e as relações interraciais no Brasil," *Revista Lugar Comum*, no. 29 February 2012. http://uninomade.net/wp-content/files_mf/110610120215A ideologia da miscigenação e as relações interraciais no Brasil - Otávio Velho.pdf.

Viana, Diego. "Turubulência à vista," *Valor Econômico*, December 22, 2015, the *Wall Street Journal Américas* sec. Interview with Ha-Joon Chang.

Vieira, Padre Antônio. "Sermão vigésimo sétimo do Rosário," in *Padre Antônio Vieira: Essencial*, ed. Alfredo Bosi, São Paulo: Penguin, 2011.

Villadsen, Kaspar. *State Phobia and Civil Society: The Political Legacy of Michel Foucault*. Redwood City: Stanford University Press, 2016.

Walras, Auguste. De la nature de la richesse et de l órigine de la valeur (Lausanne: Alcan, 1938).

Walras, Léon. *Abrégé des éléments d'économie pure*. Paris: Pichon, 1938.

Walras, Léon. *Éléments d'économie politique pure ou théorie de la richesse sociale*. Paris: Pichon et Durand-Auzias, 1926.

Weber, Max. "Au seil de la troisième année de guerre," in *Discours de guerre et d'après-guerre*, ed. Hinnerk Bruhns. Translated by Ostiane Coureau and Pierre de Larminat (Paris: Editions EHESS, 2015).

Wolf, Martin. "Os *holdouts* denigrem os abutres," *Valor Econômico*, September 3, 2014.

Xing, Li. "Introduction: Understanding the Hegemony and the Dialectics of the Emerging World Order," in Xing, *BRICS and Beyond*.

Zamora, Daniel ed. *Critiquer Foucault: Les années 1980 et la tentation néolibérale*. Bruxelles: Aden, 2014.

Zamora, Daniel, "Can We Criticize Foucault?," *Jacobin*, October 12, 2014, https://www.jacobinmag.com/2014/12/foucault-interview.

Zavaleta, René Mercado, *Lo nacional-popular en Bolivia*, first published by Siglo XXI in 1986. La Paz: Plural 2008.

Zhow, Yu. "Making Silicon Valley in China: Zhongguancun," in *The Inside Story of China's High-tech Industry: Making Silicon Valley in Beijing*, Yu Zou. Lanham: Rowman & Littlefield Publishers: 2008.

Zibechi, Raúl. "Hacer balance del progresismo," *Resumen Latinoamericano*, August 4, 2015, accessed August 2016, http://www.resumenlatinoamericano.org/2015/08/04/hacer-balance-del-progresismo/.

Zibecchi, Raúl. *Brasil potencia. Entre la integración y un nuovo imperialismo*, Malaga: Zambra - Baladre, 2012.

Žižek, Slavoj. Introduction to *Mao: On Practice and Contradiction*, by Mao Tse-Tung. London-New York: Verso, 2007.

———. "How to begin from the beginning," in *The Idea of Communism*, by Costas Douzinas and Slavoj Žižek. London: Verso, 2010.

———. *Living in the End Times*. London: Verso, 2011. 5.

———. *Organs without Bodies*. New York: Routledge, 2016.

———. *The Courage of Hopelessness: Chronicles of a Year of Acting Dangerousely*, London: Allen Lane, 2017.

Index

About the Authors

Bruno Cava graduated in aeronautic engineering at the Instituto Tecnológico de Aeronáutica (ITA), São José dos Campos, Brazil, and in law at the Universidade do Estado do Rio de Janeiro (UERJ), where he has obtained his master's degree in philosophy of law. He has published six books between politics and philosophy, analyzing urban struggles, new social movements and production of subjectivity. The author is also a regular blogger and publishes on several sites and in magazines, with articles edited in *Open Democracy*, *The Guardian*, *Le Monde Diplomatique*, *Al Jazeera*, *revue Multitudes*, *Chimères*, and *South Atlantic Quarterly*, among others. Currently, he works as an associate researcher in the Universidade Nômade network (uninomade.net) and gives free lectures in some cultural institutions at Rio de Janeiro, like the Museum of Modern Art, Casa de Rui Barbosa and Museum of the Republic.

Bruno Cava authored *A multidão foi ao deserto* (2013), which received a good review from intellectuals like Caetano Veloso and Paulo Arantes, and was later translated into Spanish in 2016. He coauthored with his fellow researcher Alexandre Mendes two books: *A vida dos direitos: ensaio sobre modernidade em Foucault e Agamben* (2008) and *A Constituição do Comum* (2017), and organized three collections of articles: with Giuseppe Cocco: *Amanhã vai ser maior* (2014), Sandra Arencón Beltrán: *Podemos e Syriza na borda das lutas* (2015), and *Márcio Pereira: A terra treme* (2016). He has had more than 30 articles accepted in scientific publications since 2007.

Some of Bruno Cava's publications:

https://www.opendemocracy.net/author/bruno-cava (English and Spanish)

https://www.theguardian.com/profile/bruno-cava (English)

http://diplomatique.org.br/o-continente-desconhecido-da-esquerda/ (Portuguese)
https://www.alfabeta2.it/tag/bruno-cava/ (Italian)
http://www.aljazeera.com/indepth/opinion/2013/12/favelados-uprising-truth-about-brazilian-protests-20131251344395l0760.html (English)
http://www.cairn-int.info/publications-of-Cava-Bruno--53676.htm (French)
https://www.diagonalperiodico.net/tags-autores/bruno-cava (Spanish)

Giuseppe Cocco is a political theorist, professor of the Universidade Federal do Rio de Janeiro, Brazil. He obtained his doctorate in social history at Université de Paris I (Panthéon-Sorbonne) (1993). He has published extensively on political violence, urban violence, the transformations of capitalism, the transformations of labor, urban territories and urban planning, citizenship, and social mobilization. He is part of the *Multitudes* collective in France and he is an editor of the social and political theory journal *Lugar Comum* in Brazil. He is also one of the founders of *Universidade Nômade Brasil*, a platform for analysis of the contemporary political context by political theorists, social theorists, and philosophers. He was made a visiting professor of the Department of Psychosocial Studies at Birkbeck University College of London between March 2017 and February 2018.

Cocco coauthored with Antonio Negri the book *GlobAL* (2007)—with its Portuguese version *Glob(AL): Biopoder e lutas em uma América Latina globalizada* (2005). Two of his recent books are *Mundobraz: o devir-Brasil do mundo e o devir-mundo do Brasil* (2009) and *KorpoBraz:Por uma política dos corpos* (2014). He edited the Special Issue "The Insurgent Multitude in Brazil," *The South Atlantic Quaterly* 113, no. 4 (Fall 2014) and has coedited (with Barbara Szaniecki) *Creative Capitalism, Multitudinous Creativity* (Lexington Books, 2015).

Cocco is a key figure of Rio de Janeiro activism and a public intellectual, having contributed in numerous and significant ways to the public debate in Brazil in the past two decades.

Some of Giuseppe Cocco's publications in *Multitudes*: http://www.cairn.info/publications-de-Cocco-Giuseppe--14147.htm.

One interview with Giuseppe Cocco, on the Brazilian uprising: https://www.viewpointmag.com/2013/06/28/brazilian-revolt-an-interview-with-giuseppe-cocco/.

Vacarme: http://www.vacarme.org/article2898.html.

www.ingramcontent.com/pod-product-compliance
Ingram Content Group UK Ltd.
Pitfield, Milton Keynes, MK11 3LW, UK
UKHW012037150926
475560UK00001B/57

9 781498 526661